Quality Management

Theory and Application

Quality Management

Theory and Application

PETER D. MAUCH

CRC Press
Taylor & Francis Group
Boca Raton   London   New York

CRC Press is an imprint of the
Taylor & Francis Group, an **informa** business

CRC Press
Taylor & Francis Group
6000 Broken Sound Parkway NW, Suite 300
Boca Raton, FL 33487-2742

Printed in the United States of America on acid-free paper
10 9 8 7 6 5 4 3 2 1

International Standard Book Number: 978-1-4398-1380-5 (Hardback)

Library of Congress Cataloging-in-Publication Data

Mauch, Peter D., 1954-
    Quality management : theory and application / Peter D. Mauch.
        p. cm.
    Includes bibliographical references and index.
    ISBN 978-1-4398-1380-5 (hardcover : alk. paper)
    1. Total quality management. 2. Quality control. I. Title.

HD62.15.M377 2009
658.4'013--dc22   2009042044

Visit the Taylor & Francis Web site at
http://www.taylorandfrancis.com

and the CRC Press Web site at
http://www.crcpress.com

Dedication

This book is dedicated to those quality professionals, past, present, and future, who seek to advance the quality sciences through traditional education and scientific research. This book is also dedicated to my close friends and family members, who gave and did not take, built and did not destroy. Especially to my wife Julie, who believed in me.

Epigraph

THE THINKER

Back of the beating hammer
By which the steel is wrought,
Back of the workshop's clamor
The seeker may find the Thought—
The Thought that is ever master
Of iron and steam and steel,
That rises above disaster
And tramples it under heel!
The drudge may fret and tinker
Or labor with lusty blows,
But back of him stands the Thinker,
The clear-eyed man who knows;
For into each plow or saber,
Each piece and part and whole,
Must go the Brains of Labor,
Which gives the work a soul!
Back of the motors humming,
Back of the belts that sing,
Back of the hammers drumming,
Back of the cranes that swing,
There is the eye which scans them
Watching through stress and strain,
There is the Mind which plans them—
Back of the brawn, the Brain!
Might of the roaring boiler,
Force of the engine's thrust,
Strength of the sweating toiler—
Greatly in these we trust.
But back of them stands the Schemer,
The thinker who drives things through;
Back of the Job—the Dreamer
Who's making the dreams come true!

Berton Braley

Contents

List of Figures

List of Tables

Preface

The rise of the quality profession as a specialty within business coincided with the increased complexity of business enterprises. In simpler times, when goods and services were provided by individual artisans, elaborate quality systems were unnecessary. An individual producer could simply compare customer requirements to his or her work and estimate its value.

The rise of complex and large enterprises produced the need for the development of objective and equitable quality procedures for the determination of value so the owners could assure the efficiency of their operations. Traditional quality management concerned itself with developing procedures to determine product conformance or nonconformance (inspection).

Unfortunately, in too many cases, the study of (inspection) procedures became an end in itself. Businesses lost sight of the objectives of the procedures. "Acceptable" techniques were applied whether they were appropriate or not. This in turn led to criticism of quality management as a discipline that provided a great deal of largely irrelevant data to management.

Fortunately, the discipline is changing. Quality professionals are becoming much more concerned with providing information that will help management meet the firm's goals. In this book, I hope to continue the movement toward consideration of the objectives of quality management and value reporting.

In quality management, fairness and objectivity play an almost equal role with relevance in the determination of the appropriate quality procedures. The American National Standards Institute (ANSI) specifies activities that must be followed as generally accepted quality principles and practices (GAQP) known as ISO 9000. In some cases, the opinions of groups with regard to these practices are backed only by the accepted stature of the promulgating organizations, whereas in others, the imposed requirements carry the weight of law behind them. In nearly every case, the intent of the suggested (or required) practice is to promote fairness in quality and the reporting of quality performance information to a diverse management audience.

A quality management system (QMS) is a *performance-reporting system* and is defined as a formal system of accumulating and reporting data useful for the achievement of management's objectives. Whether we are concerned with a not-for-profit institution or any other organization, there are general characteristics that the performance-reporting system must possess. In the following chapters, we will explore the implementation and application of a quality management system.

1

Organizing for Quality

OBJECTIVES

1. To introduce the quality organization function.
2. To discuss the quality management delegation process.
3. To present different quality organizational structures.

TERMINOLOGY

Attribute: A characteristic inherent in or ascribed to something.

Authority: The right to command and expend resources.

Category: A group of similar classifications that contain the same (multiple) attributes.

Centralized organization: An organization in which little or no authority is delegated.

Classification: A group of items in a category that contain the same (single) attribute.

Management: A process or form of work that involves the guidance or direction of a group of people toward organizational objectives, goals, or requirements.

Organization: People working together in groups to attain objectives.

Organizing: Categorizing and classifying activities, under a manager, necessary to attain objectives.

Performance measurement: The use of statistical evidence to determine progress toward specifically defined organizational objectives, goals, or standards.

Quality: Meeting customer needs.

Responsibility: Accountability for obtainment of an objective through the utilization of resources and adherence to policies.

Span of control: The number of subordinates that can be effectively managed.

Variable: A continuous or discrete measurable factor, characteristic, or attribute of an item, process, or system; some factor that might be expected to vary over time or between objects.

INTRODUCTION

The purpose of organizing is to establish lines of authority. A line of appropriate authority creates order within the company. This is necessary in order to prevent chaos where everybody is trying to do everything at once. To create synergism, departments and individuals need to work together in a coordinated effort resulting in higher efficiency. In effect, three people working together can do more work than ten people working separately. Another benefit of organizing the business is more efficient communication and reduced conflict by ensuring that authority and responsibility coincide.

CATEGORIZING DUTIES

Organizing can be viewed as categorizing activities in a business by some meaningful attributes. In most cases, business activities can be categorized into *leadership, product or service producing*, and *support*. A category is a responsibility center with an activity or collection of activities controlled by a single individual. In the quality organization and planning process, objectives are proposed for each responsibility center. The responsibility center then becomes the focal point for planning and control.

Leadership

The leadership category comprises those individuals who provide direction and guidance within the company. This includes establishing policies, goals, objectives, and standards. This group has authority and responsibility throughout the organization, from the overall system down to its individual processes. In general, the highest level of leadership (executive management) in the company is responsible for the overall direction (objectives) of the business. Individual process leaders (departmental managers) in turn are responsible for the procedures needed to achieve a given objective. Each individual with decision-making authority in an organization has responsibility for some aspect of achieving the company's objectives. It is essential to recognize this through the development of the quality management system. That is, the focus of the performance-reporting system is on responsibility centers.

Product or Service Producing

A product- or service-producing center (see Figure 1.1) is also called a *product realization center* if the person responsible has authority for producing or providing products or services to the customer.

The product- or service-producing category contains those individuals who are directly engaged in providing an output to the customer. This encompasses sales, engineering, and production activities. Ironically, the perceived order of importance of these activities is in reverse order. That is, production activities are the most visible, whereas engineering and sales are indirectly perceived as impacting the overall output. The opposite is

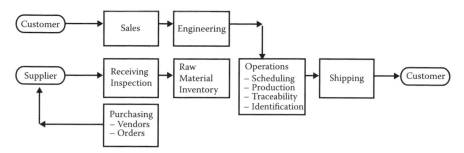

FIGURE 1.1
Product- or service-producing activities.

actually true. If the need for the product or service was not defined properly, the error will be propagated to the engineering and production functions. The same is true for poor product or service designs.

Support

A *support center* (see Figure 1.2) is a category in which the manager has authority only for providing management information or internal services within the organization with regard to product realization activities' efficiency and effectiveness.

Individuals involved in providing the leadership group with information used to make decisions regarding the efficiency and effectiveness of the business are considered support staff. This would include accounting, human resources, and quality. These individuals provide information and resources with regard to business performance. This can be in the form of a financial report, employee training, or efficiency reports.

Incongruence

Product- or service-producing and support categories cannot be intermixed. If a support activity is wrongfully placed into a product or service function, it produces group incongruence resulting in negative entropy. Eventually the magnitude of the incongruence can become so great that it may result in system (business) failure. For example, having finance reports written under sales would cause a conflict that would create chaos because they have diametrically opposed purposes. The same would be true if you placed the quality group under production.

Business Organization and Planning	**Business Improvement**
Quality Management System Management Responsibility Resource Management	Equipment Calibration Measurement, Analysis and Improvement

FIGURE 1.2
Support activities.

BREAKING CATEGORIES INTO CLASSIFICATIONS

Leadership Classifications

The structure of the classifications under the leadership category can vary slightly from company to company. The main component of these classifications revolves around the principle of the *span of control*. This refers to the number of subordinates a manager can effectively manage. The number of people who should report directly to any one person should be based upon the complexity, variety, and proximity of the work.

In practice, this turns out to be a ratio of 1:5 or 1:7. Therefore, every five to seven workers would report to a lead person; in turn, five to seven leads would report to a supervisor (5^n); and so on. This would mean that a supervisor would be able to effectively manage twenty-five workers; a manager could lead 125, and a director 625 in each department or group. The higher levels of management (directors and managers) should be spending the majority of their time organizing and planning the activities in their respective departments, while the lower levels of management (leads and supervisors) are predominantly involved in worker motivation and control. Examples of managerial-level classifications are as follows:

- President
- Vice president
- Director
- Manager
- Supervisor
- Lead

Product- or Service-Producing Classifications

The product- or service-producing categories' associated classifications are those activities that are related to providing product to the customer. This includes transformation processes where raw material is turned into finished goods. In some cases, this may mean writing a sales order, scheduling production, issuing a purchasing document to purchase raw materials, or manufacturing a product. Examples of product realization classifications are as follows:

- Customer-related processes (sales)
- Design control (engineering)
- Purchasing (production)
- Customer-supplied property (production)
- Product identification and traceability (production)
- Process control (production)
- Preservation of product (shipping and receiving)
- Servicing

Support Classifications

A support center is a category in which the manager has authority only for providing management information or internal services, with regard to a product realization center's efficiency and effectiveness. Their related classifications are associated with the administration of the business. In some cases, this may mean writing business performance statements or financial statements, or providing training to employees. Examples of support classifications are as follows:

- Control of documents
- Monitoring and measurement of product
- Monitoring and measurement of processes
- Calibration
- Control of nonconforming product
- Corrective and preventative action
- Control of quality records
- Internal quality audits
- Human resources
- Finance (accounting)

BASIC FUNCTIONAL STRUCTURE

From Figure 1.3, you can see the basic line and staff organizational structure. The organizational departments are defined by the nature of the work they perform. *President* refers to the individual who establishes the broad company policies, objectives, goals, and standards. It is expected that the individuals in the leadership group provide monthly or weekly reports to

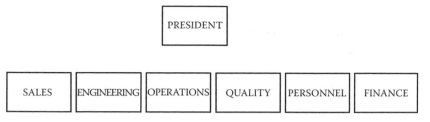

FIGURE 1.3
Line and staff organization.

the support functions with respect to the outputs of their departments showing progress toward goal attainment. An example of this would be an expense report given to accounting, training records to human resources, or yield reports to quality.

Sales refers to the function that defines product features, product promotion (including inside and outside sales), distribution (product market placement), and product pricing. The engineering function designs a product based upon product features and pricing, among other things. The operations function is responsible for reproducing the design in quantities necessary to meet customer demand. Collectively, these are the product- or service-producing functions.

The quality, accounting (finance), and human resources (personnel) functions perform support activities and provide management with information and reports with respect to company efficiency and effectiveness. The accounting department issues a monthly report called a *profit and loss statement* (P&L statement) showing financial effectiveness. The quality department issues a business quality report (BQR) that shows the performance of the organization in meeting customer needs, whereas the human resources group issues a human factor utilization report. Combined, they provide a picture of the health of the organization. For example, the P&L statement only reports the results of financial transactions, whereas the BQR shows the performance of the transactions. Collectively, these are the support functions.

Quality Function Considerations

The role of the quality department is to identify, analyze, summarize, and report the efficiency of business operations in meeting customer requirements. Additionally, the quality department may be called upon to manage certain quality projects to improve the efficiency of business operations. The need to understand the role of the quality department is an essential step in effective utilization of its resources. The following are some basic assertions:

1. Company management is responsible for quality (the financial and operational efficiency is senior management's responsibility).
2. Company management cannot delegate responsibility for quality (business efficiency starts and stays at the top); it is not a bottom-up process.
3. The quality department is not responsible for ensuring quality (senior management must take actions with regard to operational efficiency, i.e., allocate resources, goals, and performance appraisals).
4. The quality department's activities are similar to those of accounting: they only report performance, and it is up to management to act. For example, if sales are off, you would not reproach the accounting group.
5. Quality management encompasses the entire business, not just one department.

AUTHORITY, ACCOUNTABILITY, AND RESPONSIBILITY

Table 1.1 shows a responsibility matrix that clearly defines operational authority, accountability, and responsibility. The first column identifies

TABLE 1.1

Responsibility Matrix

		3. Responsibility	
1. Categories and Classifications	**2. Department**	**Primary**	**Alternate**
Product or Service			
Customer-related processes	Sales	Tom	Alice
Design control	Engineering	Mary	Jim
Purchasing	Production	Sue	Alice
Customer-supplied property	Production	Sue	Alice
Product identification and traceability	Production	Mary	Alice
Process control	Production	Mary	Alice
Preservation of product	Shipping and receiving	Hal	Sam
Servicing	Quality	Sally	Andy

the organization's functional categories and associated classifications. The next column identifies the department responsible for performing and managing the classified activities. The last column identifies specific individual responsibility (leadership). These individuals are accountable for the performance of their identified functional areas. It is senior management's responsibility to hold these individuals accountable for their respective areas. These individuals have the right to command and expend resources in their functional areas.

AUTHORITY PRINCIPLES

Delegation: There is little debate about delegation of authority. When management successfully delegates, their time is freed to pursue more important tasks, and subordinates gain feelings of belonging and added. This produces genuine feelings of commitment by workers and is the best method for development.

Unity of command: Workers should have one and only one immediate supervisor. The difficulties in servicing more than one supervisor have been well established for the last 3,000 years. In fact, almost 30 percent of all personnel problems can be related to disunity.

Scalar: The authority in an organization flows one link at a time, through the various management links. This is based upon the need for communications; circumventing the process may cause pertinent and vital information to be missed.

Exception: This principle states that managers should concentrate their efforts on matters which deviate from the norm and should allow subordinates to handle routine matters. It is believed that abnormal issues require more of the manager's abilities. Additionally, this prevents managers from becoming bogged down in routine tasks.

REVISE AND ADJUST

Due to the organic nature of the organizational structure, it should be reviewed and revised as the complexity of the company changes.

From time to time, senior management may need to add or delete a classification, department, or individual responsibility. These changes should not occur more than once per year. To do this any more frequently would be evidence of a dysfunctional organization.

COMMUNICATION

Once the organizational process has been completed by senior management, it must be published and enforced. There should be no overlapping responsibilities or departmental incongruence with respect to functional categories and classifications. The responsibility matrix in Table 1.1 shows the internal communication path for specific roles and responsibilities. For example, for sales questions you would talk to Tom or Alice, not Sally.

SUMMARY

Paying attention to the specific categorization and classification of tasks performed, and grouping those into departments should prevent incongruence and chaos in the business. This process is vital because it forms the foundation and basis for planning. There is no sense in establishing short- or long-term plans when departmental and individual managerial roles have not been properly defined. Distinguishing between product-producing and support groups should be a priority. Mixing the two would cause negative entropy and, in a worst-case scenario, business failure due to a highly dysfunctional operation.

REVIEW QUESTIONS

1. What is the purpose of organization?
2. Describe the different categories in the organization.
3. Define the leadership category.
4. Define the basic functional structure of a business.

5. Describe organizational incongruence.
6. Describe how individual accountability, responsibility, and authority are identified.
7. Describe the principles of authority.
8. Describe the role of the quality department.
9. Define the support group.
10. Explain the following:
    A. Product or service function
    B. Support function
    C. Leadership functions

2

Planning for Quality

OBJECTIVES

1. To emphasize the importance of planning in the quality management system.
2. To compare and contrast formal and informal planning.
3. To provide a systematic approach to planning.

TERMINOLOGY

Formal plan: A written, documented plan developed through an identifiable process.

Functional plans: Plans that originate from the functional areas of an organization, such as production, sales, and personnel.

Goal: (used interchangeably with *objective*) A statement that gives the organization or its departments direction and purpose.

Long-range plans: Plans that pertain to a period of time beyond the current year.

Objective: (used interchangeably with *goal*) A statement that gives the organization or its departments direction and purpose.

Planning: A process of deciding what objectives to pursue in the future and in what order to achieve them.

Policies: Broad guidelines for action which are interrelated with goal attainment.

Procedures: A series of related tasks expressed in chronological order to achieve policies.

Short-range plans: Plans that cover the current year.

Strategic planning: Planning that covers multiple years.

Tactical planning: Planning that presupposes a set of goals handed down from upper management.

INTRODUCTION

Planning is the management function that produces and integrates objectives, strategies, and policies. The planning process answers three basic questions:

1. Where are we now?
2. Where do we want to be?
3. How can we get there from here?

Planning is concerned with future actions and decisions of management. By setting objectives and establishing a course of action, management commits to "making it happen." Planning is the easiest where change happens the least. Planning is the most *useful* where change is the greatest. Most planning is carried out on an informal basis. This occurs when management does not record their thoughts and instead carries them around

TABLE 2.1

Formal versus Informal Planning

Planning	
Formal	**Informal**
Rational	Emotional
Systematic	Disorganized
Reviewed and updated	Sporadic
Used for improvement	Mostly for show
Documented	Memory based

in their heads. Table 2.1 shows the contrast between formal and informal planning.

BUSINESS QUALITY PLANNING

Elements of an Effective Quality System

The quality management system (QMS) assumes that each group performs their intended responsibilities (see Table 2.2). A worst-case scenario would be where executive management performs the role of lower-level management, leaving the company leaderless and dysfunctional.

Marketing

Management should identify the current market position of the company: where is the customer, and how and what do they buy? This would entail the identification of the market scope and depth of the products or services being offered. Additionally, management should be able to discern the market share they hold in contrast to that of their competition. A vital step is to identify current and future customer needs in terms of product features and benefits by surveying the marketplace. This step is critical to the organization's success. The major steps in marketing are as follows:

1. Identify the customer(s). (Market)
2. Identify the customer's product and service needs. (Product features)
3. Identify how much the customer is willing to pay. (Pricing)
4. Identify where the customer goes to buy. (Placement)
5. Identify how the customer hears about companies like yours. (Promotion and sales)

Setting Objectives

Setting objectives requires a cascade approach down through the company hierarchy as follows:

TABLE 2.2

QMS Responsibility Matrix

Level	Organizing	Planning	Control	Staffing	Motivation
Executive (System)	Develops the company organizational structure	Establishes the departmental policies and objectives	Monitors summary reports showing progress toward objectives	Identifies and recruits the management staff for the various departments	Meeting overall company objectives (monthly *effectiveness*) BQR
Management (Process)	Develops departmental groups responsible for specific tasks	Develops group objectives, requirements, and procedures to achieve executive objectives	Reports and monitors group progress (output) toward objectives	Identifies and recruits associates for the various groups capable of achieving objectives	Meeting departmental objectives (weekly *efficiency*) activity report
Associate (Product)	Works within a group established by management	Regulates their tasks to ensure that work is done in a consistent manner	Records data related to the department's output requirements	Collaborates work with others in the group	Ensuring the *accuracy* of the output produced daily (recordkeeping forms)

1. It begins at the top with a clear statement of what you are in business for.
2. Long-range goals are formulated for this statement.
3. The long-range goals provide the bases for short-term objectives (they are linked).
4. Objectives are established at every relevant level and function in the company.
5. This process continues down throughout the entire company.

This goal-setting process does not imply any specific management style. It does ensure that all departments and functions are in step with the major company objectives and that there is no incongruence.

Long- and Short-Range Objectives

Long-range objectives usually extend beyond the current year. These objectives must support the organizational purpose. Short-range objectives should be derived from an analysis of the long-range objectives. The analysis should result in an establishment of priorities that apply at all the various levels in the company and are synchronized with each other and the long-range objectives. The major steps in establishing objectives are as follows:

1. Formulate long-term goals.
2. Develop overall objectives.
3. Establish departmental objectives.
4. Formulate functional quality plans.
5. Establish performance metrics.
6. Implement.
7. Review performance.

After the goals have been established, an action plan for achieving the goals should be developed as follows:

1. Determine major activities needed to meet the objectives.
2. Determine subactivities under the major activities.
3. Assign responsibility for each activity.
4. Identify resources required to meet goals.

TABLE 2.3

Business Quality Plan

1. Categories and Classifications	2. Department	3. Responsibility		4. Tracking	5. Goal or Objective
		Primary	Alternate		
Product or Service					
Customer-related processes	Sales	Tom	Alice	Sales revenue per month (invoicing)	> Last month
Design control	Engineering	Mary	Jim	Project hours and cost	Per project
Purchasing	Production	Sue	Alice	Budget returns and allowances	<= Last month < 2%
Customer-supplied property	Production	Sue	Alice	Inventory	Zero spoilage
Product identification and traceability	Production	Mary	Alice	N/A	N/A
Process control	Production	Mary	Alice	Rates and yields	
Direct cost	Per quote				
Preservation of product	Shipping and receiving	Hal	Sam	Back orders/On-time delivery	<= 5%/95%
Servicing	Quality	Sally	Andy	Time and material	Per quote

The first three steps were identified in the responsibility matrix shown in Table 2.2 under business organization. Establishing and allocating the goals are discussed next.

SETTING BUSINESS METRICS

The process of determining objectives and goals is directly related to the functional categories (see Table 2.2) in the business, starting with the product- or service-producing activities. These activities are critical for the survival of the organization, where nonconformities have an immediate impact on cash flow. Some of these activities are revenue centers, while others are cost centers. The objective should be established to maximize revenue and/or reduce (or control) cost.

In Table 2.3, sales would be a revenue center (*we take money in*), where we would try to establish a realistic maximum goal or objective. Purchasing, on the other hand, is a cost center (*we pay money out*), where we would want to establish a realistic minimum (or control) goal or objective. It is senior management's responsibility to find the optimum balance between revenue, cost, and expected market share while setting objectives. There is always a cost associated with operating a business, and it is unrealistic to assume there isn't. Therefore, fixed costs based upon functional area throughput should always be considered a normal part of the process.

The very nature of the strategy- and goal-setting process is dynamic and interactive. For the most part, we would be tracking actual results and comparing them to the plan (actual/plan) in order to determine our progress and performance. The results of the current goals may change and lead to a revised strategy.

PROCESS QUALITY PLANNING

Each of the classifications or subactivities can be further analyzed and planned for their respective requirements. This is done by identifying the process tasks in chronological order. In short, this is a task listing without any of the detail. Detailing each task or step would require an explanation of how each step is accomplished. However, in process planning we only

TABLE 2.4

Process Quality Plan: General Information

General Information			
No.100	Classification (Process): Sales		Date xx/xx/xxxx
Phase: ☐Design ☐Review ☑Production	Contact Name: PDM		Phone: 999-9999 x999
Department: Sales	Primary: Tom		Alternate: Alice
Tracking: Sales revenue per month (invoicing)		Goal: > Last month	

need to know what steps or tasks are performed, not the actual "how-to" information. An example of this is shown in Table 2.4.

General Information

Table 2.4 shows the general information section of the process quality plan (PQP). It is derived from the business quality plan shown in Table 2.2. Each process plan is assigned a unique number for cross-reference and identification purposes. Then the classification and process step is identified, along with the effective date of the plan. The planning phase is identified; *design* is where the plan is in the process of development, *review* is when the plan is awaiting approval, and *production* is when the plan is in effect. Additionally, the plan identifies the departmental responsibilities, including performance tracking and goal(s). This last step is important, because this information will be used to update and revise the plan as necessary. It will also be used for the establishment of the performance measurement system and control.

Details

The details of the PQP begin with a simple procedural analysis flowchart (Table 2.5, column 1). Procedural analysis flowcharts are a useful means of making a "step-by-step" analysis of processes. The details of present (or proposed) procedures can be recorded, which will help point out duplications of effort, time delays, excessive inspection, and transportation. Analysis of existing systems can stimulate an analysis of major process changes. Adjacent to each symbol, each task is described with a title (column 2). Next to each description, we would identify any product or process requirement (column 3) as shown in Table 2.5.

TABLE 2.5

Process Quality Planning: Partial Detail

1. Flowchart					2. Process Step Description	3. Requirement (Product or Process)
Operation	Transportation	Inspection	Delay	Storage		
●	○	○	○	○	Select next sales order.	Oldest date
○	○	●	○	○	Check salesperson's math.	Correct price
○	●	○	○	○	Walk to accounts receivable file.	N/A
●	○	○	○	○	Find customer's balance.	Name and account number
●	○	○	○	○	Record customer's balance.	Correct amount

Failure Modes

Since there are no perfect processes, it will be necessary to identify problems early to control or eliminate them from happening. To do this we must determine what possible problems we may encounter during each process step (see Table 2.6). Of course, through our experience or from performing experiments, we can deduce the cause of these problems. Whenever possible, we should design the process in such a way as to reduce or eliminate all possible problems. Realistically, the elimination of all problems is not possible, but we can reduce their impact and have contingencies for their occurrence. This brings up a point: *why do we have problems?* Usually the reason there are problems in a process is because it was designed that way. If the process was put together ad hoc and informally, the output will be erratic. Couple this with inconsistent or poor management leadership, and it is truly amazing that any work gets accomplished.

The planning process provides consistency in purpose and direction of action to the accomplishment of departmental goals. In short, a good quality management system rewards actions, not words. This is the vital

TABLE 2.6

Process Quality Planning: Failure Modes

1. Flowchart					2. Process Step Description	3. Requirement (Product or Process)	4. Possible Problems	5. Possible Causes
Operation	Transportation	Inspection	Delay	Storage				
●	○	○	○	○	Select next sales order.	Oldest date	Orders mixed up	Salesperson rushed
○	○	●	○	○	Check salesperson's math.	Correct price	Wrong amount	Salesperson calculated wrong
○	●	○	○	○	Walk to accounts receivable file.	N/A	N/A	N/A
●	○	○	○	○	Find customer's balance.	Name and account number	File not found	Salesperson failed to identify new customer
●	○	○	○	○	Record customer's balance.	Correct amount	Balance wrong	Calculation incorrect

difference between a professional business manager and a novice. A novice relies upon hearsay, heightens unimportant issues, and makes decisions on gut feelings and emotions. A professional relies on information derived from statistical analysis with regard to the process for which he or she has responsibility. As you can see from Table 2.6, careful consideration is given in the design of a process to make it foolproof against error. If you want to keep problems away, you had better plan.

Control

The final step in process quality planning is to determine the internal controls for process stability. In Chapter 3, we will take an in-depth look at control systems. For our purposes here, we will explore how errors are sensed. In Table 2.7, we begin to describe how deficiencies are detected (call sensors). Errors can be detected directly or indirectly. Those that can be directly sensed are done so by making observations of the object, or error in this case. This can be done visually by looking at it or through a test instrument applied to the object. Test instruments may include rulers, micrometers, calipers,

TABLE 2.7

Process Quality Planning: Control

1. Flowchart					2. Process Step Description	3. Requirement (Product or Process)	4. Possible Problems	5. Possible Causes	6. Sensor	7. Methods		8. Document	9. Reaction Plan
Operation	Transportation	Inspection	Delay	Storage						Sample	Frequency		
●	○	○	○	○	Select next sales order.	Oldest date	Orders mixed up	Salesperson rushed	Visual	1	*	Sales order	Call manager.
○	○	●	○	○	Check salesperson's math.	Correct price	Wrong amount	Salesperson calculated wrong	Calculator	1	*	Sales order	Call manager.
○	●	○	○	○	Walk to accounts receivable file.	N/A	N/A	N/A	N/A	N/A	N/A	N/A	N/A
●	○	○	○	○	Find customer's balance.	Name and account number	File not found	Salesperson failed to identify new customer	Visual	1	*	File	Call manager.
●	○	○	○	○	Record customer's balance.	Correct amount	Balance wrong	Calculation incorrect	Calculator	1	*	File	Call manager.

* = All.

gauges, microscopes, viscosity tubes or cups, thermocouples, odometers, and hydrometers, to name a few. All these devices make direct measurements on objects. The measurements can be compared to requirements to determine if the process is operating within defined limits.

Indirect measurements are those which monitor the effect of an object. Test instruments for indirect measurement are multimeters, oscilloscopes, volt meters, amp meters, air speed gauges, gravimetric meters, dosimeters, and spectrum analyzers, to name a few. These devices measure the effects of the objects they measure and the object itself.

In Table 2.7, we describe the measurement method (sensor) along with the sample size and frequency. Both of these are derived statistically. Also included are provisions for the identification of any procedures or records used in the task. Last, when errors do occur, the last column describes what steps to take to remediate the problem.

PROJECT PLANNING

While the process quality plan is established for those activities which are an integral part of the business, there are those times when planning occurs for short-duration processes. When this occurs, we must apply the concepts of *project* planning. A project plan (see Table 2.8) is made up of significant events or milestones that must occur in some time sequence in order for a project to be completed. A project plan is a schedule of tasks over the duration of the project. Project plans are an effective means of depicting a project schedule and reporting progress as it occurs. The type of plan most often used is a Gantt-type chart, as shown in Table 2.8. When viewing this plan, you should remember that the responsible manager should have a list of all the projects under his or her control along with their associated status. In turn, each project on the list is then delegated to a project manager.

General Information

The general information section of the plan again identifies the responsible parties for the project, along with associated departmental responsibility and accountability. This helps ensure organizational integrity, line of communication, and structure.

TABLE 2.8

Project Planning: Gantt Chart

General Information

No. 100	Project Name: Demo		
Project Manager: PDM			Date 99/99/9999
Department: Engineering	Primary: Tom	Alternate: Alice	Phone: 999-9999 x999
Planned Hours: 100	Actual Hours: 50	AI: 0.5	
Planned Cost: $3,750	Actual Cost: $1,850	CI: 0.49	
Start Date: 99/99/9999	Stop Date: 99/99/9999	Status Date: 99/99/9999	SI:1.02

1. Activity or Document	2. % Completed	3. Status	4. Period Ending (Week)												
			1	2	3	4	5	6	7	8	9	10	11	12
Study phase	45	●												
Initial market analysis	100	●												
Product scope and depth	100	●												
Team feasibility report	50	○												
Feature listing	0	○												
Functional requirement	0	○												
Capability report	0	○												

Note: ● = completed, ○ = incomplete,

Status Reporting

The next few blocks on the project plan in Table 2.8 are status blocks, which are used to track the project's progress as well as to record the planned and actual, time and cost. The Achievement Index (AI) is calculated by dividing the actual hours by the planned hours. An AI value of less than 1.00 represents underachievement. Accordingly, the Cost Index (CI) is calculated by dividing the actual cost (time and material) by the planned cost. A CI value greater than 1.00 represents overexpenditure. The overall Status Index (SI) is calculated by dividing the AI by the CI. An SI between .9 and 1.1 is normal; greater 1.3 or less than .7 would require immediate attention.

Detail

In the body of the Gantt chart, the first column is used to identify the tasks to be performed. This may require you to break down a task into constituent parts called *parent-child relationships*. This can be seen in Table 2.8, where the *study phase* task is the parent and the steps below are the children belonging to this task. Adjacent to each step is a reporting column for percentage completed, where we would designate what percentage of the step has been finished. Next to this is a column which provides a graphical status indicator. The plot portion of the chart shows a bar for the duration of the task or step in which black represents completion and gray indicates the scheduled time allotted. The bar turns black as the project progresses based upon the percentage completed. From Table 2.9, you can see that the *feature listing* step has not been started, even though it was scheduled to start in week 7.

PRODUCT QUALITY PLANNING

General Information

At last, we have come to the actual product or service itself. Product quality planning (see Table 2.9) is by no means the last step; in fact, it starts when the product is being designed. This should be part of the design test phase outputs prior to the actual design testing. Typically, this plan is developed conjointly with the product illustrations (e.g., drawings and schematics), bill of materials, and production work order. These plans are utilized in raw

TABLE 2.9

Product Quality Plan

Subject:				Effective Date:	Number: 99999
Part Number: 999-9999 Revision A				Supersedes: 99/99//99	Page: 9 of 9
Approved By: ZZZZZZ					

Instructions:

Inspect the product to the characteristics listed below (also see drawing and/or inspection and test work instructions). Use a C = 0 sampling plan with an acceptable quality level of 10 unless otherwise specified below. Record the results of the inspection on the appropriate inspection report or log. In the event of a nonconformity, follow work instructions.

No.	Characteristics to Be Measured or Inspected	Specification and Tolerance (±)		Acceptable Quality Level	Inspection or Measuring Equipment or Method	Comments
①	Overall length	12"	.25	1.0	Caliper	
②	Inside diameter hole *A*	.25"	.005		Micrometer	
③	Inside diameter hole *B*	.25"	.005		Micrometer	
④	Hole *A* location	1.5"	.010		Caliper	
⑤	Hole *B* location	2.5"	.010		Caliper	
⑥	Overall width	6"	.250		Caliper	
⑦	Thickness	.250"	.005		Caliper	
⑧	Overall height	3.0"	.025		Caliper	
⑨	Color: tan	—	—		Visual	
⑩						

material, work-in-process components, and finished goods. These plans are used to perform product verification and validation. In some cases, these are called *test plans* or *specifications*. From Table 2.9, you can see there is a general information area called *subject* where part information is entered, as well as room for effective dates and approval and control numbers. Additionally, there is an area for special instructions where necessary.

Detail

In the body of Table 2.9 is a column called *No.*, which is used to count the rows on the form and also serves to reference a characteristic on the product illustrations (e.g., drawings or schematics). This number can be annotated on the illustration to correspond to the characteristics being specified in the product quality plan. Adjacent to each number is a description of the characteristic to be measured, which can be either a discrete or continuous variable. For continuous characteristics, we need to identify a target value with an upper and lower limit.

There may be occasions where you would need to specify an acceptable quality level (AQL) for a particular characteristic. The term *AQL* refers to the percentage nonconforming and is used for determining an appropriate sample size. Since not all characteristics are created equally, some are more critical than others. Those with lesser criticality may be assigned a higher AQL than those that are more critical.

For each characteristic, we should identify the type of measurement method or equipment to use. This will help inspection and test planning as well as help determine the competency requirements for personnel performing the tests. It also provides congruence with suppliers of raw material with regard to testing methods. Of course, you could also identify a documented test procedure in this column. Last, there is a column for making any additional comments.

PRODUCT VERIFICATION AND VALIDATION PLANNING

Responsibility and Interfaces

Responsibility within the company for the validation planning should be established. As part of validation planning, responsibilities for validation activities and functions for supporting and interfacing departments should be determined. Typically, supporting and interfacing departments include manufacturing, engineering, purchasing, and others. Arrangements for coordination with other validation groups or departments should also be identified.

Information Accessibility

Project information, such as contracts, schedules, work orders, specifications, drawings, manuals, procedures, configuration of operating equipment, and purchase orders, should be available to the personnel who plan for validation. Requests for quotes and bid proposals may be obtained if they contain information useful to the validation-planning function.

Files and Records

Facilities and files for maintaining forms, including tags, hard copy, and computers, should be available. Capability for obtaining and maintaining relevant specifications, drawings, contracts, and other documents in readily accessible files should be established. Capability for validation records storage and protection for established retention periods and retrieval from files should also be established. Additionally, there should be access to referenced documents such as standards (such as those of the American National Standards Institute [ANSI] or ASTM International [ASTM]) that set forth acceptance criteria, texts on validation sampling, and other pertinent documents, including applicable codes.

Validation Facilities

Facilities and validation equipment required for performing validations should be determined and provided as necessary. Consider some of the following:

1. Space and equipment (surface plates, tools, gauges, etc.) for validation, affording environmental conditions (e.g., appropriate lighting level, temperature and humidity control, and cleanliness level) consistent with handling and validation needs of products and services
2. Facilities for receiving and handling items being validated (shelving, storage areas)
3. Facilities for taking verification samples and for containing test results to validate significant characteristics
4. Facilities for maintenance of archived samples where critical materials are involved
5. Physical validation systems and equipment for performing validation and testing, including dimensional, electrical, mechanical, and pneumatic examination; nondestructive examination (NDE); and destructive examination (DE)
6. Facilities or alternate provisions for calibration of measuring equipment

Validation Personnel

The validation-planning system should consider the availability of product analysts with the capability (education and technical training) to perform the types of validations required. Typically, broad-based capabilities for dimensional, optical, nondestructive, and destructive evaluation and testing provide the greatest versatility.

Validation Procedures

Product analysts should be provided detailed guidelines, checklists, instructions, or procedures when necessary to supplement the drawings, specifications, and other applicable documents.

Scheduling and Revising Validation Plans

The validation-planning process should include a process for determining the need for validation plans and for initiating such plans. These should be developed in conjunction with the manufacturing and construction operation process plans.

POLICIES, PROCEDURES, AND OBJECTIVES

Organizational goals, policies, and procedures are not mutually exclusive components (see Figure 2.1). Each is related to the other; for example, policies relate to objectives, while procedures relate to policies. Similarly, they are an integral part of the organizational structure. Policies identify what departments do, whereas procedures tell us how to do it. Goals are achieved through policies and procedures.

In and of themselves, they can do nothing unless senior management is dedicated to making them happen. There are many cases where companies have failed to follow the correct course of action. This is due in large part to senior management becoming self-focused, where their own well-being and self-interest take priority over those of the organization. Policies, procedures, and goals then become imaginary rather than realistic. Being imaginary, the objectives become negative motivators destroying the creditability of senior management. A lack of integrity leads to a dysfunctional

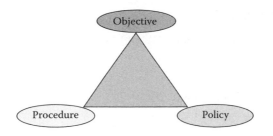

FIGURE 2.1
Interaction of objectives, policy, and procedures.

organization. In fact, any organization that operates on the premise of "What's in it for me?" will find it difficult to achieve true quality results.

Policies

The first step in establishing company policies is to identify the customer, how the customer buys, and how the customer can be reached. Second, top management must determine customer needs. Top management should determine what the present and future business should be. The next step is to establish organizational responsibility, authority, and resources as follows:

1. Determine major activities in the company.
2. Determine subactivities.
3. Assign primary and alternate responsibilities.
4. Identify the resources needed.

Policies exist at all levels of the organization. A typical organization has policies that relate to everyone in the company. Policies outline a general course (or framework) of action to be followed and do not precisely describe how to achieve specific objectives.

A general outline for writing a policy statement is as follows:

1. Describe the major activities and subactivities.
2. Identify the objective.
3. Define the department that is responsible.
4. Identify the associated procedure.
5. State the policy.

In most cases, the policy statements are grouped together into a manual (e.g., a quality policy manual). This manual is typically assigned a control number for reference, and is dated and approved by senior management.

Procedures and Rules

Procedures and rules define in step-by-step fashion the methods through which policies are achieved. They outline the manner in which a recurring activity must be accomplished. Procedures should allow for flexibility and deviation.

Rules require that specific actions be taken with respect to a given situation (step). Rules leave little doubt concerning what is to be done. They permit no flexibility or deviation. Unlike procedures, rules do not necessarily specify a sequence.

A basic outline for a procedure is as follows:

1. No. (control number assigned for reference)
2. From (the person approving the procedure)
3. To (the person responsible for executing the procedure)
4. Date (the date the procedure was approved)
5. Subject (major activity)
6. Regarding (subactivity)
7. CC: (carbon copy list)
8. Procedure steps and rules

Typically a master list is maintained that identifies all the procedures used by the organization along with the approval date.

FORMS AND RECORDS

The task of designing forms for quality is the responsibility of the quality management. The management representative must know (1) what data the user wants to collect, and (2) how the form is going to be used in the quality system.

The basic parts of a form are as follows:

1. Title (identifies the form)

2. Instructions (tells how to complete the form)
3. Heading (contains all the general data)
4. Body (specific data the form is designed to collect)
5. Conclusion (contains approvals, signatures, and summary data)

There are two basic styles of forms: open style and boxed style. The open style is the simplest. It consists of headings and open areas in which data can be collected. The boxed style allocates space to each data item. Each box is clearly identified by name or by a brief description. Forms are seldom purely "open" or purely "boxed." They are usually described as predominantly open or boxed, or as a combination of both. Completed forms are considered records.

Typical control of the forms requires a number to be assigned to the form. This number is usually found in the footer section of the form. A date is commonly found next to this number signifying the date the form was placed into use. These forms are then listed on the document master list. The storage location, filing method, and retention times must be identified for completed forms (or records).

BLUEPRINTS (PRODUCT SPECIFICATIONS)

Product specifications developed internally should be identified along with a revision letter code which can be cross-referenced in some manner to a definition of what changed. Normally, there is a list showing the history of revisions for each part. In most cases, these specifications are maintained in a filing system.

PROCESS FLOWCHARTING

Flowcharting is a graphical technique specifically developed for use in computer science. It is a pictorial representation that uses predefined symbols to describe data flow in a business, or the logic of a computer program or process. The symbols shown in Table 2.10 are "predefined"; their shapes identify data and communicate what is happening to the data.

TABLE 2.10

Flowcharting Symbols

Symbol	Description
	Terminal symbol indicates the start, stop, halt, pause, or interruption in a process.
	Process symbol is a representation of a task performed in the processes.
	Decision symbol is used for operations that determine which of two or more alternative paths will be followed in a process.
	On-page connector is used to connect or link other flowchart symbols.
	Off-page connector is used when the flowchart is continued on another page.
	Document symbol is used to describe any input or output that is a paper document.

Flowcharts help quality professionals to describe and communicate complex sets of processes and data in three principle ways:

1. Analyze existing processes.
2. Synthesize new processes.
3. Communicate with others.

Standardization of Symbols

National and international efforts to develop standard symbols began in the early 1960s. The efforts in the United States resulted in the set of symbols adopted by the ANSI. See Table 2.10.

Normal Logic Flow

The normal logic flow is downward and by columns from left to right.

Process Flow

As shown in Table 2.11, there are three basic process flows that have been identified in processes: (1) sequential, (2) loop, and (3) if-then-else. These flows appear the most often when one is describing processes. Your process can be a combination of any of these. As you will notice, for the most part each symbol has one input and one output, except in the case of the decision symbol, where there are two outputs (true or false).

TABLE 2.11

Basic Process Flows

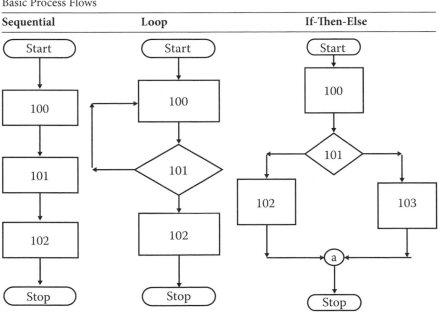

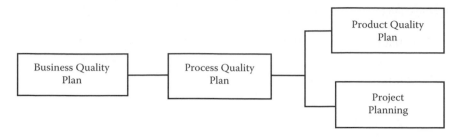

FIGURE 2.2
The quality-planning process.

COMMUNICATION

The business quality plan in Table 2.2 is the first plan to be developed. The process quality plan in Table 2.6 is derived from the business quality plan, ideally for each activity listed. For short-term processes a project plan (Table 2.8) is used, while a product quality plan is made for each type of material used from raw material to finished goods. These plans can then be revised as needed. The axiom "Measure twice and cut once" more than applies when it comes to planning. These plans are communicated from the senior management down throughout the entire organization and provide the basis for meaningful dialogue within the company.

SUMMARY

From Figure 2.2, you can see that the planning process starts with the business quality plan, from which in turn the process quality plans are derived. Additionally, product and project plans are generated based upon the process quality plan. Projects can be in the form of design control for new product development, or the implementation of corrective or preventive actions. The product quality plans define the features, functions, and characteristics of the product and/or its components. These plans are integral to each other and cannot be performed separately. Nor can these plans start with the product and work backward. Doing so would indicate a lack of management commitment and overdelegation.

REVIEW QUESTIONS

1. What is the purpose of organizational planning?
2. Describe the different types of plans.
3. Define the business quality plan.
4. Define the basic process and product plans.
5. Describe goal incongruence.
6. Describe how the various plans relate to each other.
7. Describe the principles of project planning.
8. Describe how goals are defined.
9. Define how plans are communicated.
10. Explain the following:
    A. The business-planning function
    B. The process-planning function
    C. The product- and project-planning functions

3

Controlling for Quality

OBJECTIVES

1. Define the control process, and discuss the elements of reporting.
2. Develop an appreciation for business, process, and product performance reporting.
3. Introduce and describe the various reporting structures.

TERMINOLOGY

Defect: A departure of a quality characteristic from its intended level or state that occurs with a severity sufficient to cause an associated product or service not to satisfy intended normal, or foreseeable, usage requirements.

Management: A process or form of work that involves the guidance or direction of a group of people toward organizational goals or objectives.

Management (quality) control: A process for setting goals, monitoring performance, and correcting for deviations.

Nonconformity: A departure of a quality characteristic from its intended level or state that occurs with a severity sufficient to cause an associated product or service not to meet a specification requirement.

Objective: A statement (used interchangeably with *goal*) designed to give an organization and its members direction and purpose.

Policies: Broad, general guidelines for action which relate to goal attainment.

Procedures: A series of related steps or tasks expressed in chronological order to achieve a specific purpose.

Process approach to management: An approach to the study of management that focuses on the management functions of planning, controlling, organizing, staffing, and motivating.

Rules or requirements: Guidelines that require specific and definite actions be taken with respect to a given situation or task.

Systems approach to management: A philosophy, popularized by Frederick Taylor, concerning the relationship between people and work that seeks to increase productivity and simultaneously make work easier by scientifically studying work methods and establishing standards rather than depending on tradition and custom.

INTRODUCTION

Defining the organizational purpose is critical with regard to control. Management must identify the customer, where the customer is, how the customer buys, and how the customer can be reached. Next, management must determine what the customer buys.

In addition to defining the present business, management must also identify what future business will be and what it should be.

Long-Range Objectives

Long-range objectives generally extend beyond the fiscal year of the organization. Long-range objectives must support and not be in conflict with the stated organizational purpose. However, long-range objectives may be quite different from the organizational purpose and still support it.

Short-Range Objectives

Short-range objectives should be derived from in-depth evaluation of long-range objectives. Such an evaluation should result in a listing of priorities of the long-range objectives. Once the priorities have been established, short-range objectives can be set to help achieve the long-range objectives.

A Cascade Approach

One approach to setting objectives is to have the objectives "cascade" down through the organizational hierarchy. The objective-setting process begins at the top with a clear, concise statement of the central purpose of the organization. Long-range organizational goals are formulated from this statement. The long-range goals lead to the establishment of more short-range performance objectives for the organization. Derivative objectives are then developed for each major division or department. Objectives are then established for the various subunits in each major division or department. The process continues on down through the organization.

The following items represent potential areas for establishing objectives in most organizations:

1. Profitability
2. Markets
3. Productivity
4. Product
5. Financial resources
6. Physical facilities
7. Research and innovation
8. Organization
9. Human resources
10. Customer service

Concerns of Control

In order to maintain stability, the manager must be sure that the organization is operating within its established boundaries of constraint. The next concern is objective realization, which requires continual monitoring to ensure that adequate progress is being made toward the accomplishment of established objectives.

At top management levels, a problem occurs whenever the organization's objectives are not being met. At middle and lower levels of management, a problem occurs whenever the objectives for which the manager is responsible are not being met. All forms of management control are designed to provide the manager with information regarding progress. Once the manager has this information, it can be used for several purposes:

1. To prevent crises
2. To standardize output
3. To appraise employee performance
4. To update plans
5. To protect an organization's assets

Written Reports

There are two basic types of written reports, analytical and informational. Analytical reports interpret the facts they present. Informational reports only present the facts.

Preparing Reports

Preparing a report is a four- or five-step process depending on whether it is informational or analytical.

1. Planning the attack
2. Collecting the facts
3. Organizing the facts
4. Interpreting the facts (analytical only)
5. Writing the report

Correcting for Deviations

All too often, managers set standards and monitor results but do not follow up with appropriate corrective actions. The first two steps are of little value if corrective action is not taken. The steps for effective corrective actions are as follows:

1. Identify the problem.
2. Perform an investigation to determine the cause of the problem.
3. State the cause of the problem.
4. Determine a solution for the cause and implement it.
5. Prove the solution removed the cause.

Importance of Value

Persons conducting a business do so with the expectation of increasing value. To accomplish this, they use their outputs to produce goods and services that are demanded in the marketplace. They sell their products and services and, in return, use outputs to produce more goods and services. To be successful, they must sell their goods or services to their customers for an amount greater than the cost of producing them. To do this, they must ensure the value of the outputs produced. The relationship of the value to the owner's outputs may be expressed by the following equation:

Value = Outputs – Nonconformance

Most businesspeople find that they are unable to conduct their business satisfactorily using only financial measures.

BASIC CONCEPTS

An organization should have a set of objectives. Management of an organization will require information for determining how well these objectives have been achieved.

This focuses on two factors:

1. Whether the goals have been met (effectiveness)
2. Whether they were able to provide products and services with minimal nonconformities (efficiency)

In general, the purpose in measuring value is to help management control the activities of the firm.

ORGANIZATIONAL RESPONSIBILITY

Each individual with decision-making authority in an organization has responsibility for some aspect of achieving his or her company's objectives. It is essential to recognize this through the development of the performance-logging system. That is, the focus of the value-logging system is on responsibility centers. A responsibility center is an activity or collection of activities controlled by a single individual. In the quality-planning process, objectives are proposed for each responsibility center. The responsibility center then becomes the focal point for control.

The type of responsibility the person in charge can exert classifies responsibility centers. A center is a *primary* or *product-producing center* if the person responsible has authority only for producing or providing products or services to the customer. In some cases, this may mean writing a sales order, issuing a purchasing document to purchase raw materials, or producing a product. Examples of primary centers are as follows:

1. Contract review
2. Design control
3. Purchasing
4. Customer-supplied product
5. Product identification and traceability
6. Process control
7. Handling, packaging, storage, preservation, and delivery
8. Servicing

A *support center* is a center in which the manager has authority only for providing management information or internal services, with regard to primary centers' efficiency and effectiveness, within the organization. In some cases, this may mean writing business performance statements or financial statements, or providing training to employees. Examples of support centers are as follows:

1. Document control
2. Inspection and testing
3. Control of inspection, and measuring and testing equipment
4. Inspection and test status
5. Control of nonconforming product
6. Corrective and preventative action
7. Control of quality records
8. Internal quality audits
9. Training

In measuring value, a distinction is made between the performance of a responsibility center and the performance of its manager. For decisions concerning whether the organization should continue to provide a product or service, all outputs and nonconformance for a responsibility center are accumulated. This practice allows management to make such decisions as whether the company should continue the process or not.

In contrast, only a subset of activity outputs and nonconformance is accumulated for measuring managerial performance. Only those nonconformities over which the manager can exert influence are included in the log for managerial performance. Managerial performance is then measured in part by comparing levels of controllable nonconformance against management's objectives. In this way, a manager may be judged to have performed efficiently for a given activity.

THE ROLE OF QUALITY MANAGEMENT

The role of the quality professional has changed dramatically from the days when the quality control manager was simply responsible for the inspection of product. Quality managers are now largely responsible for preparing detailed performance statements; they are asked to help in measuring the effectiveness of operations and suggesting improvements; and they are involved in identifying and proposing solutions to emerging problems. Quality professionals are primarily responsible for designing the firm's performance information system and assuring compliance with quality-logging requirements.

In light of the expanding duties involved, the importance of the quality function is usually recognized in a firm's organizational chart by having the quality executive report directly to the president.

INTRODUCTION TO QUALITY-REPORTING BASICS

To be useful, performance logging must be assembled and logged objectively. Those who must rely on such information have a right to be assured that the data are free from bias and inconsistency, whether deliberate or not. For this reason, performance-logging systems rely on certain standards or guides that have proved useful over the years in impart-ing valued information. These standards (ISO 9000) are called *generally accepted quality principles* (GAQP). Because quality is more an art than a science, these principles are not immutable laws like those in the physi-cal sciences. Instead, they are guides to action and may change over time. Sometimes specific principles must be altered or new principles must be formulated to fit circumstances or changes in business practices.

Because quality principles are based on a combination of theory and practice, there has always been, and probably always will be, some contro-versy about their propriety.

A principle report resulting from the process of accumulating quality information is the *business performance* or *quality report*. The business quality report portrays the operating results of primary and support cen-ter activities for a period of time. This log is prepared monthly.

Another basic log is called an *activity report* or *log* and is generally required in logging responsibility center activities. This log will be dis-cussed later.

Quality professionals log company transactions, which are the result of an activity in relationship to the amount of nonconformities, to determine performance.

THE BUSINESS QUALITY REPORT

The business quality report (also known as a *performance report*) is a list-ing of the firm's primary and support responsibility centers' activities on a given date (see Figure 3.1).

The body of the statement contains four major sections: center activities, number of outputs (assets) generated, amount of nonconformance (nc), and calculated value (percentage).

> *Outputs* (or assets) are the resources of the business that can be expressed as the output of an activity. Outputs can take many forms. Some out-puts may have readily identifiable characteristics. Others may simply represent information used to communicate customer requirements throughout the organization. Examples are sales orders, purchasing

Red Bead Company Performance Report January 30, 20XX			
Primary Centers	**Assests**	**Nc**	**Value (%)**
RAW MATERIAL INVENTORY			
4.6 Purchasing			
Vendor Assessment	150	12	92%
Purchasing Data	300	5	98%
WORK IN PROCESS	450	17	96%
4.3 Contract Review	500	3	99%
4.4 Design Control	25	2	92%
Department A	25000	1500	94%
Department B	12000	150	99%
Department C	5000	250	95%
	42000	1900	95%
FINISHED GOODS INVENTORY			
4.15 Delivery	250	11	96%

FIGURE 3.1
Business quality report.

documents, equipment calibration records, and nonconformance logs. Outputs are usually recorded at the acquisition.

Nonconformities are departures of an asset from its intended requirement, or a state that occurs with a severity sufficient to cause the asset not to satisfy customer requirements.

Value is the result of comparing outputs to nonconformance to determine percent performance for a given activity. Value is calculated as follows:

$$\left(1 - \frac{\sum Nonconformances}{\sum Assets}\right) \times 100$$

Business performance is the collective value of all the activities within a responsibility center. Figure 3.2 shows the different product-producing departments' scope of performance and where they all intersect. This intersection is where all activities are in 100 percent coincidence with each other. This means that each function provided exactly what the other function required in order to have a usable output. This is the optimum performance area of the business. The larger the

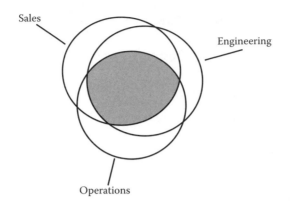

FIGURE 3.2
Performance diagram.

intersection, the better the performance. If all these functions completely overlapped, the organization would be 100 percent efficient.
In practice, this is rarely the case; studies have shown that there is an
inherent 3–7 percent error rate no matter what you do.

$$Performance = (S \cup E \cup O)$$

Example: From the Red Bead Company report, overall primary performance is calculated as follows:

$$Pp = (.99 \times .92 \times .96 \times .95 \times .96) = .80 \text{ (or 80 percent).}$$

UNDERLYING CONCEPTS

Certain fundamental concepts provide a framework for recording and
logging performance. These concepts have been developed over time to
provide general guidelines for making business quality reports as objective and as useful as possible.

Any business is an individual unit, separate and distinct from other
activities. A separate business quality report would be maintained for each
separate business. Outputs are recorded and logged on activity reports or
logs to provide a "historical record" of events. Performance activities and

their results appearing in business quality reports are expressed in terms of units generated.

A performance activity is a business activity that requires quality recognition. Therefore, an event that affects any of the elements in the value equation (outputs or nonconformance) must be logged.

ACTIVITY REPORTING

The ultimate objective of performance logging is to record the correct number of outputs generated by an activity and the amount of nonconformance. However, for practical reasons there is a very important quality-logging function, preceding the recording of outputs and nonconformance. The outputs and nonconformance are first entered in a preliminary record called an *activity report* or *log*. The process of recording outputs and nonconformance is called *journalizing*. After the outputs and nonconformance have been journalized, they are totaled and posted to the performance report or log.

Some of the reasons for activity reporting or logging are as follows:

1. It provides a chronological record of all outputs and nonconformance generated by an activity.
2. It fulfills the need for recording in one place all outputs and nonconformance for a responsibility center.
3. It provides more information, such as a detailed explanation of nonconformance, than can conveniently be recorded on individual records.

The original sources of information concerning most outputs and nonconformance are quality records. Examples are sales orders, purchasing documents, production logs, and various other types of business papers. Such records are called *source documents*. Information obtained from source documents helps in determining the number of outputs generated and amount of nonconformance found. The source documents also provide valuable evidence to support the accuracy of the performance log.

The purpose of the activity report or log is to provide a chronological record of all outputs and nonconformance for a given quality activity. It includes the date in which the outputs and nonconformance were generated, a description of the nonconformance, and the source document from which the information was collected. The flow of information for performance logging is as follows.

As shown in Figure 3.3, the source documents (quality records) provide the basic information concerning the outputs and nonconformance of each responsibility center activity. The activity report or log represents the first point where data are formally entered in the performance-logging system. Various activity reports or logs may be used, depending on the size of the firm and the nature of their operations.

The simplest form used is a multicolumn form, called an *activity report* or *log* (see Figure 3.4). It has columns for the date on which outputs were generated, asset description, total outputs generated, total nonconformance, description of nonconformance, and totals for each nonconformance description. The basic form for the activity report or log is illustrated as follows:

> *Department block.* The name of the department generating the activity report or log is entered into the department block. This is the responsibility center for the control of quality.
>
> *Activity block.* The name of the activity being logged is entered into the activity block, such as *subcontractor assessment* or *production department.*
>
> *Record used.* The name of the source document from which data were taken is entered into the record used block, such as *purchase order, sale order,* or *inspection log.*

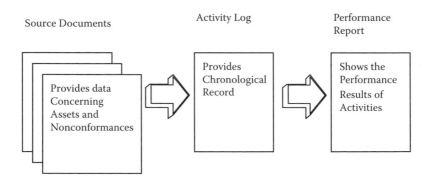

FIGURE 3.3
Reporting flow.

ACTIVITY LOG   **Page:**

Department:			Description of Nonconformance		
Activity:					
Record Used:					
Date:	Description	Total	Nc			

FIGURE 3.4
Basic activity report.

Date column. The date the asset was generated is entered in the date column. The date is entered on the first row for each asset activity.

Description column. The description column is used to record the asset activity or an explanation of the activity.

Total column. The total number (or sample size taken) of outputs generated by the activity is entered in the total column.

Nc column. The total amount of nonconformance found in the outputs is entered in the Nc column.

Description of nonconformance. Descriptions of the nonconformance are entered in the columns below the title *description of nonconformance.*

Nonconformance columns. The quantities of the total nonconformance that are attributed to the nonconformance are entered under each nonconformance description.

JOURNALIZING PROCEDURE

To illustrate the recording of activities in the activity report or log, assume that on May 1, 20xx, John Doe of the Red Bead Company Sales Department took 500 orders from customers. Upon review of these orders, three nonconformities were found. This activity is recorded on the first page of the activity report or log as follows:

ACTIVITY REPORT   **Page: 1**

Department:	Sales			Description of Nonconformance		
Activity:	Order Entry			Wrong Price	Wrong Part Number	
Record Used:	Sale Orders					
Date:	Description	Total	Nc			
5/1/xx	Sales Taken	500	3	2	1	

The recording of the outputs (sales taken) and nonconformance (including description of the nonconformance) provides a summary of each day's activities, which is a valuable reference if there is some future question about the activity.

For example, the Red Bead Company completed the following additional activities during May.

> **Activity**: May 2, 20xx, the Purchasing department wrote 150 purchase orders for raw material purchases and 12 nonconformities were found.

ACTIVITY LOG   **Page: 1**

Department:	Production			Description of Nonconformance		
Activity:	Purchasing			Wrong Quantity Ordered	Wrong Part Number	
Record Used:	Purchase Orders					
Date:	Description	Total	Nc			
5/1/xx	Orders Made	150	12	10	2	

> **Activity**: May 3, 20xx, Department A produced 25,000 units and 1,500 nonconformities were found.

ACTIVITY LOG   **Page: 1**

Department:	Production		Description of Nonconformance		
Activity:	Department A		Wrong Color (Blue)	Out of Tolerance > .05"	
Record Used:	Production Work Orders				
Date:	Description	Total	Nc		
5/1/xx	P/N 12345	25000	1500	1200	300

Activity: May 4, 20xx, Department B made 12,000 units and 150 non-conformities were found.

ACTIVITY LOG   **Page: 1**

Department:	Production		Description of Nonconformance		
Activity:	Department B		Wrong Size <.001"	Out of Tolerance > .005"	
Record Used:	Production Work Orders				
Date:	Description	Total	Nc		
5/1/xx	P/N 536	12000	150	125	25

Activity: May 10, 20xx, Department C produced 5,000 units with 250 nonconformities.

ACTIVITY LOG   **Page: 1**

Department:	Production		Description of Nonconformance		
Activity:	Department C				
Record Used:	Production Work Orders				
Date:	Description	Total	Nc	Cracked	Fading
5/1/xx	P/N 345	5000	250	175	75

Activity: May 12, 20xx, Shipping & Receiving shipped 250 units and 11 shipments were found nonconforming.

ACTIVITY LOG   **Page: 1**

Department:	Shipping & Receiving			Description of Nonconformance		
Activity:	Delivery			Raw Mat'l Delay'd	Date Change Made	
Record Used:	Packing Slips					
Date:	Description	Total	Nc			
5/1/xx	Shipments	250	11	10	1	

In the activities recorded above, there was only one entry. Many business activities require the use of more than one entry. Any entry on the activity report or log requiring more than one entry is added to the next available line until the record is complete. These logs are usually completed by the responsibility center manager and given to the quality organization for posting in the performance log.

POSTING

The process by which the outputs and nonconformance are summarized and transferred to activities on the performance log is called *posting*. It consists of transferring totals for the outputs (total) column and the nonconformance (nc) column on the individual activity reports or logs to the performance log. This is usually performed once a month by the quality organization.

In most businesses, the posting is done either manually or by data-processing methods. The conversion to paperless information systems has become very common among businesses. Posting is illustrated in Figure 3.5.

Activity Report

Department:	Sales			Description of Nonconformance		
Activity:	Order Entry			Wrong Price	Wrong Part Number	
Record Used:	Sale Order					
Date:	Description	Total	Nc			
5/1/××	Sales Taken	500	3	2	1	
	Totals	500	3	2	1	

Primary centers	Assets	Nc	Value (%)
4.3 Contract Review	500	3	99%
4.4 Design Control	25	2	92%
4.6 Purchasing			
Vendor Assesment	150	12	
Purchasing Data	300	5	
	450	17	96%
4.9 Process Control			
Department A	25000	1500	
Department B	12000	150	
Department C	5000	250	
	42000	1900	95%
4.15 Delivery	250	11	96%
Primary Performance			80%

Red Bead Company Performance Report May 30, 20××

FIGURE 3.5
Posting to the business quality report.

PRODUCT PERFORMANCE REPORTING

Inspection records should be maintained that are identifiable to batch, lot, serial number, or work order number. When data are to be taken, define the number of data points and specify the appropriate format for recording the data. An activity report or log is maintained providing a chronology of the inspections performed.

The inspection records should contain the following:

1. Inspection data and results compared to acceptance criteria. An acceptance decision should be reached for each inspection identifying whether or not compliance with acceptance criteria has been achieved. This acceptance or rejection decision should be recorded, dated, signed, and made available for review.
2. Written disposition, waivers, or deviations from authorized organizations (e.g., a material review board) for release of nonconforming items and services.
3. Inspection data, including descriptions of any nonconforming items, provided on a timely basis to responsible functions as feedback to be used for trend analysis or process improvement.

ANALYSIS

Ranked Order Analysis

The business quality report and activity reports provide a basis for establishing frequency distributions. The nonconformities become class bounties, and the numbers of nonconformities are the cumulative frequency of occurrence. The nonconforming class can then be arranged with the highest-occurring nonconformity at the top on down to the lowest, as shown in Table 3.1.

From Table 3.1, it should be obvious that *cracked* is the most frequently occurring nonconformity. Hence, this should be the first problem to be corrected.

TABLE 3.1

Ranked Order Analysis

Rank	Description of Nonconformity	Tally	f	%
1	Cracked	IIIII	5	.33
2	Loosened	IIII	4	.27
3	Leaking	III	3	.20
4	Sticking	II	2	.13
5	Fractured	I	1	.07
Summary			15	1.00

Fishbone Diagram

The fishbone diagram shown in Figure 3.6 is used to determine causes and effects which relate to a nonconformity. The major branches of the fishbone diagram are used to determine the major influences that would result in the problem outcome. The braches are as follows:

1. *Method*: This is the branch where you would describe those factors in the process that would affect the output with regard to the current practices.
2. *Machine*: Here you would identify those factors which are related to the equipment used in the process.
3. *Material*: Here you would identify those factors related to the material used in the process.
4. *People*: Here you would identify those factors related to the people working within the process.

This technique has proven to be useful in determining the causes of problems. It is interesting to note here that changing any variable in the process will result in a change in the output regardless of whether it is the actual cause.

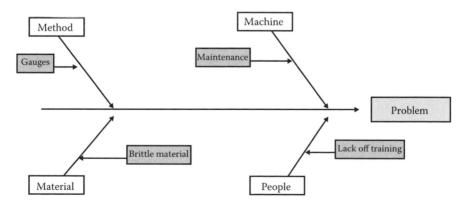

FIGURE 3.6
Fishbone diagram.

CONTROLLING NONCONFORMANCE IDENTIFICATION

Identification during Work

When operators are engaged in an activity to produce a product, the act of regulating or making modifications is considered a component of that activity. In essence, any changes made during the manufacture of a product are not considered a nonconformity.

Completion of Work

Nonconformities are usually detected after the completion of work. This occurs when the completed product (which is the result of work) is inspected with regard to preestablished requirements.

Inspection

The purpose of inspection with respect to products is as follows:

1. To provide a basis for action for the product already on hand
2. To decide if the product meets requirements
3. To provide a basis for action with regard to the process
4. To decide if the process requires action

Identification of Nonconforming Product

When in the course of inspection (by the operator or independent inspector) nonconforming product is detected, it should be identified by some suitable means (e.g., tags, markings, or location).

SEGREGATION

Short-Run Production

There are cases where the lot or batch integrity must be maintained. In such circumstances, the nonconforming units must be appropriately identified to ensure against their use. Where integrity of the lot is not an issue, units should be isolated or removed from the flow of production to prevent unintended use.

Long-Run Production

In the case of mass production (or continuous, homogeneous flow), non-conforming units must be removed from the process for disposition. Nonconforming units must be separated from other materials to prevent unintended use.

Risk

Associated nonconformance risk levels are defined as follows:

1. *Critical*: A nonconformity that may cause bodily harm, injury, or death, and/or prevents the product from performing its intended function.
2. *Major*: A nonconformity that may reduce the life of the product and/ or is readily noticeable by the customer.
3. *Minor*: A nonconformity that is neither critical nor major.

DISPOSITION

Responsibilities for Disposition

Over the years, there have been many ways to determine how to dispose of nonconforming material; the following are some current practices:

1. *Material review boards*: In some cases, organizations have opted to establish a group of top managers to review nonconforming material for disposition. This requires the consensus of all the managers in the disposition decision.
2. *Management*: Other organizations have delegated the responsibility for disposition of nonconforming product to a specific departmental manager or managers.

3. *Operators*: In some cases, due to the nature of the noncon- formity, the operator can dispose of nonconforming material immediately.

CORRECTIVE AND PREVENTIVE ACTION (CAPA) METHODOLOGY

Immediate Action Required

When the nonconformity has an associated risk of critical or major, correc- tive action should be taken immediately. If the nonconformity is detected while the product is in transit to the customer or storage, the customer must be notified at once and the material placed in quarantine.

Magnitude of the Nonconformity

An analysis should be made of the nonconformities by product to detect statistical trends. This is done using ranked order analysis (see Table 3.1) to determine the most frequently occurring nonconformity for corrective action.

CAPA Methodology

When performing either corrective or preventive action, the steps are the same. It should be noted that corrective actions are always product-based, whereas preventive actions are process- or system-based. The steps are as follows:

1. Investigate the cause of nonconformities relating to the product, pro- cess, and quality system, and record the results of the investigation.
2. Identify the root cause of issues requiring corrective or preven- tive action.
3. Determine the steps needed to deal with any problems requiring corrective or preventive action.
4. Implement corrective and preventive actions.

5. Apply controls to ensure that corrective and preventive action is taken.
6. Initiate controls to ensure that actions taken are effective.
7. Confirm that relevant information on actions taken is submitted for management review.

SUMMARY

Departmental managers are primarily concerned with the efficiency of certain activities in their department which are indicators of their overall performance. By monitoring these activities, departmental managers can prevent, reduce, or eliminate nonconformities, resulting in improved overall value. To do this, they must determine the output of each activity and its interrelationship with regard to providing value to the customer. Some activities provide support to achieving the desired output, while others are directly related to or incorporated into the output itself.

Each activity usually has a source document related to the output. This could be a purchase order, packing list, or training record. The documents provide a record of the event in writing, which preserves the knowledge of the transaction. In this way, key data are collected to provide useful information to management.

Once the data have been collected on records, these records must be categorized and summarized to provide a chronology of events related to the department's output. This chronology usually provides the following information:

1. Number of outputs produced
2. Number of nonconformities found
3. Description of the nonconformities

The departmental managers should establish goals and requirements in writing which are consistent with the overall company performance. In addition, departmental managers should identify the record used to collect pertinent data.

REVIEW QUESTIONS

1. What is the purpose of the business quality report?
2. Describe the different types of centers.
3. Define the business quality report.
4. Define the basic activity-reporting process.
5. Describe how overall business performance is calculated.
6. Describe how value is calculated.
7. Describe the three principles of quality control.
8. Describe how goals are defined.
9. Define how the reports are communicated.
10. Explain the following:
    A. The business-reporting structure
    B. Journalizing and posting
    C. The product- and project-planning function

4

Staffing for Quality

OBJECTIVES

1. Define the employee-forecasting process, and discuss the elements of resource planning.
2. Develop a scheme for the development of job descriptions and requirements.
3. Describe the various education and training methodologies.

TERMINOLOGY

Education: The act or process of imparting or acquiring knowledge, skill, or judgment.

Experience: The application of education.

Research: Research is a human activity based on intellectual investigation and aimed at discovering, interpreting, and revising human knowledge on different aspects of the world. Research can use the scientific method, but need not do so. Scientific research relies on the application of the scientific method, a harnessing of curiosity. This research provides scientific information and theories for the explanation of the nature and the properties of humans. It makes practical applications possible. Scientific research is funded by public authorities, by charitable organizations, and by private groups, including many companies. Scientific research can be subdivided into different classifications.

Speculation: Contemplation or consideration of a subject; meditation. A conclusion, opinion, or fact reached by conjecture. Reasoning based on inconclusive evidence; conjecture or supposition. Engagement in risky business transactions on the chance of quick and/or considerable profit. A commercial or financial transaction involving speculation.

Theory: A set of statements or principles devised to explain a group of facts or phenomena, especially one that has been repeatedly tested or is widely accepted and can be used to make predictions about natural phenomena. The branch of a science or art consisting of its explanatory statements, accepted principles, and methods of analysis, as opposed to practice: for example, a fine musician who had never studied theory.

FORECASTING HUMAN RESOURCES NEEDS

Forecasting and/or scenario analysis is a process of analyzing possible future staffing events by considering alternative possible outcomes (scenarios). The analysis is designed to allow improved decision making by allowing more complete consideration of outcomes and their implications.

For example, in economics and finance, a financial institution might attempt to forecast several possible scenarios for the economy (e.g., rapid growth, moderate growth, and slow growth), and it might also attempt to forecast financial market returns (for bonds, stocks, and cash) in each of those scenarios. The institution might consider subsets of each of the possibilities. It might further seek to determine correlations and assign probabilities to the scenarios (and subsets, if any). Then it will be in a position to consider how to distribute assets between asset types (i.e., asset allocation); the institution can also calculate the scenario-weighted expected return (this figure will indicate the overall attractiveness of the financial environment).

Depending on the complexity of the financial environment, economic and finance scenario analysis can be a demanding exercise. It can be difficult to foresee what the future holds (e.g., the actual future outcome may be entirely unexpected), that is, to foresee what the scenarios are, and to assign probabilities to them; and this is true of the general forecasts, never mind the implied financial market returns. The outcomes can be modeled mathematically and statistically (e.g., taking account of possible variability within single scenarios as well as possible relationships between scenarios).

Financial institutions can take the analysis further by relating the asset allocation that the above calculations suggest to the industry or peer group distribution of assets. In so doing, the financial institution seeks to control its business risk rather than the client's portfolio risk.

In politics or geopolitics, scenario analysis involves modeling the possible alternative paths of a social or political environment, and possibly diplomatic and war risks. For example, in the recent Iraq War, the Pentagon certainly had to model alternative possibilities that might arise in the war situation and had to position material and troops accordingly. The difficulty of such forecasting is highlighted in that case by the fact that it is arguable that the Pentagon failed to foresee the lawlessness and insecurity of the postwar situation and the level of hostility shown toward the occupying forces.

Scenario analysis can also be used to illuminate "wild cards." For example, analysis of the possibility of the earth being struck by a large celestial object (a meteor) suggests that while the probability is low, the damage inflicted would be so high that the event is much more important (threatening) than the low probability (in any one year) alone would suggest. However, this possibility is usually disregarded by organizations using scenario analysis to develop a strategic plan, since it has such overarching repercussions. In the case of personnel planning (see Table 4.1), the amount of additional resources that are needed can be estimated based upon the current staffing level and projected increase or decrease on the output of each process. Additionally, the required management staff can be estimated based upon the 1:5 to 1:7 ratio discussed in Chapter 1.

From Table 4.1, the current staffing and projected staffing calculated based upon expected process output increase at some future period in time. This is done by determining the current outputs based upon the business quality report (see Figure 3.1), organizational responsibility (see 4), and calculating the total hours worked by the associates (i.e., 2 associates × 160 hours per month = 320 hours total), then calculating the output rate (i.e., 230 output/320 total hours = 0.718 each). Using the projected

TABLE 4.1

Personnel Forecasting

1. Organizational Responsibility				2. Current Resources						3. Process Output [BQR]				4. Forecasted Resources						
Requirement	Department Responsible	Primary	Alternate	Associates	Leads	Supervisors	Managers	Directors	Vice Presidents	Current Output	Total Hours	Rate	Projected Output Increase	Additional	Associates	Leads[1:5]	Supervisors [1:5]	Managers [1:5]	Directors [1:5]	V. P. [1:5]
Customer related processes	Sales	Tom	Alice	2	1	1	1	1	-	250	320	0.78	50	0.40	2.40	0.48	0.10	0.02	0.00	0.00
Design control	Engineering	Mary	Jim	5	1	1	1	-	-	10	800	0.01	2	1.00	6.00	1.20	0.24	0.05	0.01	0.00
Purchasing	Production	Sue	Alice	3	1	1	1	-	-	500	480	1.04	25	0.15	3.15	0.63	0.13	0.03	0.01	0.00
Customer supplied property	Production	Sue	Alice	1	1	1	1	-	-	500	160	3.13	150	0.30	1.30	0.26	0.05	0.01	0.00	0.00
Identification	Production	Mary	Alice	1	1	1	1	-	-	100	160	0.63	20	0.20	1.20	0.24	0.05	0.01	0.00	0.00
Process control	Production	Mary	Alice	30	1	1	1	-	-	25,000	4,800	5.21	10,000	12.00	42.00	8.40	1.68	0.34	0.07	0.01
Preservation of product	Shipping/ receiving	Hal	Sam	2	1	1	1	-	-	5,000	320	15.63	1,500	0.60	2.60	0.52	0.10	0.02	0.00	0.00
Servicing	Quality	Sally	Andy	1	1	1	1	-	-	100	160	0.63	25	0.25	1.25	0.25	0.05	0.01	0.00	0.00
			Total	45	8	8	8	-	-					14.90	59.90	11.98	2.40	0.48	0.10	0.02

increase in output, we can calculate the expected personnel increase (i.e., 50 increase × 0.718 rate = 0.40 additional). When this addition is added to the existing number of associates, the total number of associates can be calculated (i.e., 2 current associates + 0.4 additional = 2.4 total associates required). By dividing the total number of associates by 5, we can derive the number of leads required (i.e., 2.4 associates/5 = 0.48 leads). This is done for each subsequent management level to determine the most efficient number of managerial levels required. For this case, there would be no need for a level higher than supervisory.

By comparing current and projected staffing needs at the various levels, an imbalance between associates and management becomes clear. In this case, there are far too many managerial personnel (too many cooks spoil the broth) and not enough associates. This organization will require a shift in responsibilities and staffing.

JOB DESCRIPTIONS

Job Analysis

The general purpose of job analysis is to document the requirements of a job and the work performed. Job analysis is performed as a preliminary to successive actions, including defining a job domain, writing a job description, creating performance appraisals, selection and promotion, training needs assessment, determining compensation, and organizational analysis and planning.

In the fields of human resources (HR) and industrial psychology, job analysis is often used to gather information for use in personnel selection, training, classification, and/or compensation.

The field of *vocational rehabilitation* uses job analysis to determine the physical requirements of a job to determine whether an individual who has suffered some diminished capacity is capable of performing the job with, or without, some accommodation.

Professionals developing certification exams should use job analysis (often called something slightly different, such as *task analysis*) to determine the elements of the domain which must be sampled in order to create a content-valid exam. When a job analysis is conducted for the purpose of valuing the job (i.e., determining the appropriate compensation for incumbents), this is called *job evaluation*.

Methods

There are several ways to conduct a job analysis, including interviews with incumbents and supervisors, questionnaires (structured, open-ended, or both), observation, and gathering background information such as duty statements or classification specifications. In job analysis conducted by HR professionals, it is common to use more than one of these methods.

For example, the job analysts may tour the job site and observe workers performing their jobs. During the tour, the analyst may collect materials that directly or indirectly indicate required skills (duty statements, instructions, safety manuals, quality charts, etc.).

The analyst may then meet with a group of workers or incumbents. And, finally, a survey may be administered. In these cases, job analysts typically are industrial psychologists or have been trained by, and are acting under the supervision of, an industrial psychologist.

In the context of vocational rehabilitation, the primary method is direct observation and may even include video recordings of incumbents involved in the work. It is common for such job analysts to use scales and other apparatus to collect precise measures of the amount of strength or force required for various tasks. Accurate, factual evidence of the degree of strength required for job performance is needed to justify that a disabled worker is legitimately qualified for disability status. In the United States, billions of dollars are paid to disabled workers by private insurers and the federal government (primarily through the Social Security Administration). Disability determination is, therefore, often a fairly "high-stakes" decision. Job analysts in these contexts typically come from a health occupation such as occupational or physical therapy.

Questionnaires are the most common methodology employed by certification test developers, although the content of the questionnaires (often lists of tasks that might be performed) is gathered through interviews or focus groups. Job analysts in this area typically operate under the supervision of a psychologist.

Results

Job analysis or descriptions can result in a description of common duties, or tasks, performed on the job, as well as descriptions of the knowledge, skills, abilities, and other characteristics (KSAOs) required for performing those tasks. In addition, job analysis can uncover tools and technologies commonly used on the job, working conditions (e.g., a cubicle-based environment, or outdoor work), and a variety of other aspects that characterize work performed in the position(s). When used as a precursor to personnel selection (a commonly suggested approach), job analysis should be performed in such a way as to meet the professional and legal guidelines that have been established (e.g., in the United States, the Uniform Guidelines on Employee Selection Procedures).

In certification testing, the results of the job analysis lead to a document for candidates laying out the specific areas that will be tested (named in various ways, such as *exam objectives*), and to a *content specification* for item writers and other technical members of the exam development team. The content specification outlines the specific content areas of the exam and the percentage (i.e., the number) of items that must be included on the exam from that content area.

Position Requirement

As a minimum, you should identify each title in the business through the use of the organizational chart (see Figure 1.3). Then, for each title (see Figure 4.1), you should describe the education and experience required and the tasks performed. Once the analysis is done, the various positions in the company should be staffed by competent employees as defined by the requirements.

From Figure 4.1, the department has been identified: sales in this case and one of the positions under sales. This position is further broken down into education required, experience required, and tasks. The *tasks* category has been further broken down into the individual steps. These in turn are then documented as position requirements for a given title. Note that you can have several people with the same title; therefore, the number of titles should not exceed the number of employees. In fact, there should be very few titles in the organization.

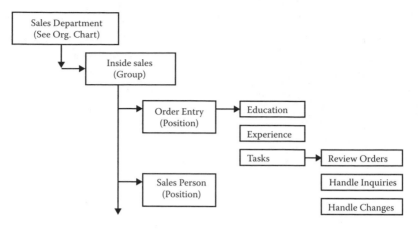

FIGURE 4.1
Organizational breakdown of positions.

EDUCATION AND TRAINING

Systems of Formal Education

Educational systems are established to provide education and training, in most cases for children and the young. A curriculum defines what students should know, understand, and be able to do as the result of education. A teaching professional delivers teaching which enables learning, and a system of policies, regulations, examinations, structures, and funding enables teachers to teach to the best of their abilities. *Education* is a broad concept; it refers to all the experiences in which people can learn something. *Instruction* refers to the intentional facilitating of learning toward identified goals, delivered either by an instructor or in other forms. *Teaching* refers to learning facilitated by a real live instructor. *Training* refers to learning that prepares learners with specific knowledge, skills, or abilities that can be applied immediately.

Primary Education

Primary (or elementary) education consists of the first years of formal, structured education. In general, primary education consists of six or seven years of schooling starting at the age of five or six, although this varies between and sometimes within countries. Globally, around

70 percent of primary-age children are enrolled in primary education, and this proportion is rising. Under the Education for All programs driven by UNESCO, most countries have committed to achieving universal enrollment in primary education by 2015, and in many countries, it is compulsory for children to receive primary education. The division between primary and secondary education is somewhat arbitrary, but it generally occurs at about eleven or twelve years of age. Some educational systems have separate middle schools, with the transition to the final stage of secondary education taking place at around the age of fourteen. In the United States and Canada, schools which provide primary education are referred to as *primary schools*. Primary schools in these countries are often subdivided into infant schools and junior schools.

Secondary Education

In most contemporary educational systems of the world, secondary education consists of the years of formal education that occur during adolescence. Secondary education is characterized by being the transition from the typically compulsory, comprehensive primary education for minors, to the optional, selective tertiary, postsecondary, or higher education (e.g., university or vocational school) for adults. Depending on the system, schools for this period or a part of it may be called *secondary* or *high schools, gymnasiums, lyceums, middle schools, colleges,* or *vocational schools*. The exact meaning of any of these terms varies between the systems. The exact boundary between primary and secondary education varies from country to country and even within them, but is generally around the seventh to the tenth year of schooling. Secondary education occurs mainly during the teenage years. In the United States and Canada, primary and secondary education together are sometimes referred to as *K–12 education*, and in New Zealand *Year 1–13* is used. The purpose of secondary education is to give common knowledge, and to prepare for higher education or to train directly in a profession.

Higher Education

Higher education, also called tertiary, third-stage, or postsecondary education, is the noncompulsory educational level following the completion of a school providing a secondary education, such as a high school, secondary school, or gymnasium. Tertiary education normally includes undergraduate and postgraduate education, as well as vocational education and training. Colleges and universities are the main institutions that provide tertiary education. Tertiary education generally results in the receipt of certificates, diplomas, or academic degrees.

Higher education includes the teaching, research, and social services activities of universities, and within the realm of teaching, it includes both the undergraduate level (sometimes referred to as *tertiary education*) and the graduate (or postgraduate) level (sometimes referred to as *graduate school*). Higher education in the United States and Canada generally involves work toward a degree-level or foundation degree qualification. In most developed countries, a high proportion of the population (up to 50 percent) now enters higher education at some time in their lives. Higher education is very important to national economies, both as a significant industry in its own right, and as a source of trained and educated personnel for the rest of the economy; for example, the following is a list of academic program levels:

1. *Associate degree*: Requires 60 credits
2. *Bachelor degree*: Requires 120 credits
3. *Master degree*: Requires 30 credits (bachelor degree or assessment equivalency required)
4. *Master of business administration*: Requires 40 credits (bachelor degree or assessment equivalency required)
5. *Doctorate*: Requires 60 credits (master's degree or assessment equivalency required)

Adult Education

Lifelong learning, or adult education, has become widespread in many countries. Adult education takes on many forms, from formal class-based learning to self-directed learning.

Alternative Education

Alternative education, also known as *nontraditional education* or *educational alternative*, is a broad term which may be used to refer to all forms of education outside of traditional education (for all age groups and levels of education). This may include both forms of education designed for students with special needs (ranging from teenage pregnancy to intellectual disability), and forms of education designed for a general audience which employ alternative educational philosophies and/or methods.

Alternatives of the latter type are often the result of education reform and are rooted in various philosophies that are commonly fundamentally different from those of traditional compulsory education. While some have strong political, scholarly, or philosophical orientations, others are more informal associations of teachers and students dissatisfied with certain aspects of traditional education. These alternatives, which include charter schools, alternative schools, independent schools, and home-based learning, vary widely but often emphasize the value of small class size, close relationships between students and teachers, and a sense of community.

List of Adult Alternative Educational Methods

1. *Degrees by assessment*: Portfolio assessment of knowledge acquired through experience and prior learning.
2. *Degrees by research*: Independent study, academic projects, and assignments.
3. *Degrees by exam*: "Testing out," passing online exams such as the College Level Examination Program (CLEP).
4. *Degrees by distance courses*: Distance learning programs offer accelerated coursework-based degrees for busy individuals who want a high-quality education without interrupting their present careers or family responsibilities.
5. *On-the-job training (OJT)*.

TRAINING

Training refers to the acquisition of knowledge, skills, and competencies as a result of the teaching of vocational or practical skills and knowledge that relate to specific competencies. It is the core of apprenticeships and provides the backbone of content at technical colleges and polytechnics.

In addition to the basic training required for a trade, occupation, or profession, observers of the labor market recognize the need to continue training beyond initial qualifications in order to maintain, upgrade, and update skills throughout a person's working life. People in many professions and occupations refer to this sort of training as *professional development*.

Some people use a similar term for workplace learning to improve performance: *training and development*. One can generally categorize such training as on-the-job or off-the-job:

1. On-the-job training takes place in a normal working situation, using the actual tools, equipment, documents, or materials that trainees will use when fully trained.
2. On-the-job training has a general reputation as most effective for vocational work.
3. Off-the-job training takes place away from normal work situations— implying that the employee does not count as a directly productive worker while such training takes place.
4. Off-the-job training has the advantage that it allows people to get away from work and concentrate more thoroughly on the training itself. This type of training has proven more effective in inculcating concepts and ideas.

Training has specific goals of improving one's capability, capacity, and performance.

Training Records

Each employee (exempt or not) should have a title. An individual training record should be maintained, as shown in Figure 4.2. The record identifies the employee by first and last name. The record also identifies the title of the individual, as well as his or her highest level of education and initial hire date. A log is provided to list all training received and the date. This information should be verified with a copy of a diploma, certificate, or school transcript which is attached to the record. Additionally, there is a column for the name or initials of the person who verified the training or education.

EMPLOYEE TRAINING RECORD

Last Name:		First Name:		Title:
Highest Educational Level Achieved:			Date Hired:	

Date	Description of Skills, Training or Education Received	Verified by

FIGURE 4.2
Employee training record.

SUMMARY

The staffing function is a direct derivative of the organizing function. The company must first establish the organizational structure and identify how activity will be carried out under planning. Once planning has been performed, then organizational control is its natural extension. During the planning and control process, we identified the tasks, tracking, and objectives that are to be met. This in turn leads to the staffing function to accomplish the company's mission. A personnel forecast is developed based upon expected process output. The forecast defines the personnel requirements at all levels in the organization. Position requirements are then established to identify education, experience, and tasks performed. Once the position requirements have been defined, the organization develops education and training requirements, including records of these activities.

REVIEW QUESTIONS

1. What is the purpose of the personnel forecast?
2. Describe the job analysis process.
3. Define how titles are identified.
4. Define the basic position requirements for a given title.
5. Describe the different levels of education.
6. Describe adult education.
7. Describe the alternative methods of education.
8. Describe on-the-job training.
9. Define different levels of postsecondary education.
10. Explain the following:
    A. The training record
    B. How training is verified
    C. Experience

5

Motivating for Quality

OBJECTIVES

1. To develop an understanding of the motivation process.
2. To identify leadership and management styles.
3. To explore how motivation impacts performance.

TERMINOLOGY

Goodwill: A kindly feeling of approval and support; benevolent interest or concern; the favor or advantage that a business has acquired, especially through its brands and its good reputation.

Happiness: A state of well-being and contentment.

Job: Something produced by or as if by work (did a nice *job*).

Leader: Person who directs a group or unit; a person who has commanding authority or influence.

Morale: The mental and emotional condition (as of enthusiasm, confidence, or loyalty) of an individual or group with regard to the function or tasks at hand.

Motivation: A compelling force, stimulus, or influence.

Satisfaction: Fulfillment of a need or want.

LEAD, COACH, AND GUIDE

Leadership is closely linked with the idea of management. The two are synonymous; management is a subset of leadership. With this premise, you can view leadership as follows:

1. Centralized or decentralized
2. Broad or focused
3. Decision-oriented or morale-centered
4. Intrinsic or derived from some authority

Any of the bipolar labels traditionally ascribed to management styles could also apply to leadership styles. Hersey and Blanchard (1982) use this approach: they claim that management merely consists of leadership applied to business situations, or in other words, management forms a subset of the broader process of leadership. They put it this way: "Leadership occurs any time one attempts to influence the behavior of an individual or group, regardless of the reason. Management is a kind of leadership in which the achievement of organizational goals is paramount."

However, a clear distinction between management and leadership may nevertheless prove useful. This would allow for a reciprocal relationship between leadership and management, implying that an effective manager should possess leadership skills, and an effective leader should demonstrate management skills. One clear distinction could provide the following definition:

1. Management involves power by position.
2. Leadership involves power by influence.

Abraham Zaleznik (1977), for example, delineated differences between leadership and management. He saw leaders as inspiring visionaries concerned about substance, while viewing managers as planners who have concerns with process. Warren Bennis (1989) further explicated a dichotomy between managers and leaders. He drew twelve distinctions between the two groups:

1. Managers administer; leaders innovate.
2. Managers ask how and when; leaders ask what and why.
3. Managers focus on systems; leaders focus on people.
4. Managers do things right; leaders do the right things.
5. Managers maintain; leaders develop.
6. Managers rely on control; leaders inspire trust.
7. Managers have a short-term perspective; leaders have a longer-term perspective.
8. Managers accept the status quo; leaders challenge the status quo.
9. Managers have an eye on the bottom line; leaders have an eye on the horizon.
10. Managers imitate; leaders originate.
11. Managers emulate the classic good soldier; leaders are their own person.
12. Managers copy; leaders show originality.

Paul Birch (1999) also sees a distinction between leadership and management. He observes that, as a broad generalization, managers concern themselves with tasks, while leaders concern themselves with people. Birch does not suggest that leaders do not focus on "the task." Indeed, the things that characterize a great leader include the fact that they achieve. Effective leaders create and sustain competitive advantage through the attainment of cost leadership, revenue leadership, time leadership, and market value leadership. Managers typically follow and realize a leader's vision. The difference lies in the leader realizing that the achievement of the task comes about through the goodwill and support of others (influence), while the manager may not.

This goodwill and support originate in the leader seeing people as people, not as another resource for deployment in support of "the task." The manager often has the role of organizing resources to get something done. People form one of these resources, and many of the worst managers treat people as just another interchangeable item. A leader has the role of motivating others to follow a path he or she has laid out, or a vision he or she has articulated in order to achieve a task. Often, people see the task as subordinate to the vision. For instance, an organization might have the overall task of generating profit, but good leaders may see profit as a by-product that flows from whatever aspect of their vision that differentiates their company from the competition.

Leadership does not only manifest itself as purely a business phenomenon. Many people can think of an inspiring leader they have encountered who has nothing whatever to do with business: a politician, an officer in the armed forces, a Scout or Guide leader, a teacher, and so on. Similarly, management does not occur only in the context of business. Again, we can think of examples of people who we have met who fill the management niche in nonbusiness organizations. Nonbusiness organizations should find it easier to articulate a non-money-driven inspiring vision that will support true leadership. However, often this does not occur.

Differences in the mix of leadership and management can define various management styles. Some management styles tend to deemphasize leadership. Included in this group, one could include participatory management, democratic management, and collaborative management styles. Other management styles, such as authoritarian management, micromanagement, and top-down management, depend more on a leader to provide direction. Note, however, that just because an organization has no single leader giving it direction, this does not mean it necessarily has weak leadership. In many cases, group leadership (multiple leaders) can prove effective. Having a single leader (as in a dictatorship) allows for quick and decisive decision making when needed, as well as when not needed. Group decision making sometimes earns the derisive label *committee-itis* because of the longer times required to make decisions, but group leadership can bring more expertise, experience, and perspectives through a democratic process.

Patricia Pitcher (1994) has challenged the division into leaders and managers. She used a factor analysis (in marketing) technique on data collected over eight years and concluded that three types of leaders exist, each with very different psychological profiles: artists were imaginative, inspiring, visionary, entrepreneurial, intuitive, daring, and emotional; craftsmen were well-balanced, steady, reasonable, sensible, predictable, and trustworthy; and technocrats were cerebral, detail-oriented, fastidious, uncompromising, and hard-headed. She speculates that no one profile offers a preferred leadership style. She claims that if we want to build, we should find an "artist leader"; if we want to solidify our position, we should find a "craftsman leader"; and if we have an ugly job that needs to get done like downsizing, we should find a "technocratic leader." Pitcher also observes that a balanced leader exhibiting all three sets of traits occurs extremely rarely: she found none in her study.

Bruce Lynn postulates a differentiation between leadership and management based on approaches to risk. Specifically, "A Leader optimizes upside opportunity; a Manager minimizes downside risk" ( ). He argues that successful executives need to apply both disciplines in a balance appropriate to the enterprise and its context. Leadership without management yields steps forward, but as many (if not more) steps backward. Management without leadership avoids any steps backward, but doesn't move forward.

Leadership Styles

Autocratic

An *autocratic* or authoritarian manager makes all the decisions, keeping the information and decision making among the senior management. Objectives and tasks are set, and the workforce is expected to do exactly as required. The communication involved with this method is mainly downward, from the leader to the subordinate. Critics such as Elton Mayo have argued that this method can lead to a decrease in motivation from the employee's point of view. The main advantage of this style is that the direction of the business will remain constant, and the decisions will all be similar; this in turn can project an image of a confident, well-managed business. On the other hand, subordinates may become highly dependent upon the leaders and increased supervision may be needed.

Paternalistic

A more *paternalistic* form is also essentially dictatorial; however, the decisions tend to be in the best interests of the employees rather than the business. A good example of this would be David Brent running the business in the British version of the fictional television show *The Office*. The leader explains most decisions to the employees and ensures that their social and leisure needs are always met. This can help balance out the lack of worker motivation caused by an autocratic management style. Feedback is again generally downward; however, feedback to the management will occur in order for the employees to be kept happy. This style can be highly advantageous, and can engender loyalty from the employees, leading to a lower labor turnover rate, thanks to the emphasis on social needs. It shares similar disadvantages to an authoritarian style though, with employees becoming highly dependent on the leader. If the wrong decisions are made, then all employees may become dissatisfied with the leader.

Democratic

In a *democratic* style, the manager allows the employees to take part in decision making; therefore, everything is agreed on by the majority. The communication is extensive in both directions (from subordinates to leaders and vice versa). This style can be particularly useful when complex decisions need to be made that require a range of specialist skills: for example, when a new information and communication technologies (ICT) system needs to be put in place and the upper management of the business is computer illiterate. From the overall business' point of view, job satisfaction and quality of work will improve. However, the decision-making process is severely slowed down, and the need for a consensus may lead to not taking the "best" decision for the business. It can go against a better choice of action.

Laissez-Faire

In a *laissez-faire* leadership style, the leader's role is peripheral and staff manages their own areas of the business; the leader therefore evades the duties of management, and uncoordinated delegation occurs. The communication in this style is horizontal, meaning that it is equal in both directions; however, very little communication occurs in comparison with other styles. The style brings out the best in highly professional and creative groups of employees; however, in many cases it is not deliberate and is simply a result of poor management. This leads to a lack of staff focus and sense of direction, which in turn leads to much dissatisfaction and a poor company image.

REWARDS BASED UPON PERFORMANCE

A psychological reward is a process that reinforces behavior—something that, when offered, causes a behavior to increase in intensity. *Reward* is an operational concept for describing the positive value an individual ascribes to an object, behavioral act, or internal physical state. Primary rewards include those that are necessary for the survival of the species, such as food, water, and sex. Some people include shelter as a primary reward. Secondary rewards derive their value from the primary rewards and include money, pleasant touch, beautiful faces, music, and the like. The functions of rewards are based directly on the modification of behavior and less directly on the physical and sensory properties of rewards. For instance, altruism may induce a larger psychological reward, although it doesn't cause physical or sensory sensations, thus favoring such behavior, also known as *psychological egoism*. Rewards are generally considered more effective than punishment in enforcing positive behavior. Rewards induce learning, approach behavior, and feelings of positive emotions.

PRAISE AND CENSURE FAIRLY

In its common usage, *praise* is the act of making positive statements about a person, object, or idea, either in public or privately. Praise is often contrasted with *criticism*, where the latter is held to mean exclusively *negative* statements made about something, although this is not technically correct. Most people are responsive to praise and their self-esteem or confidence will increase if a suitable amount of praise is received—in fact, some psychological theories hold that a person's life is composed largely of attempts to win praise for his or her actions. Other people are less affected by or even averse to praise, for example people with autism or schizoid personality disorder.

Performance Appraisals

A performance appraisal is a regular review of employee performance within organizations. Generally, the aims of such a scheme are as follows:

1. Give feedback on performance to employees in meeting group goals and objectives.
2. Identify employee training needs.
3. Document criteria used to allocate organizational rewards.
4. Form a basis for personnel decisions: salary increases, promotions, disciplinary actions, and so on.
5. Provide the opportunity for organizational diagnosis and development.

6. Facilitate communication between employee and administrator.
7. Validate selection techniques and human resources policies to meet federal Equal Employment Opportunity requirements.

A common approach to assessing performance is to use a numerical or scalar rating system whereby managers are asked to score an individual against a number of group or departmental objectives or attributes. In some companies, employees receive assessments from their manager, peers, subordinates, and customers while also performing a self-assessment.

The most popular methods that are being used as performance appraisal processes are as follows:

1. Management by objectives (MBO)
2. Behavioral Observation Scale (BOS)
3. Behaviorally Anchored Rating Scale (BARS)

Trait-based systems, which rely on factors such as integrity and conscientiousness, are also commonly used by businesses. The scientific literature on the subject provides evidence that assessing employees on factors such as these should be avoided. The reasons for this are twofold:

1. Because trait-based systems are by definition based on personality traits, they make it difficult for a manager to provide feedback that can cause positive change in employee performance. This is caused by the fact that personality dimensions are for the most part static, and while an employee can change a specific behavior, he cannot change his personality. For example, a person who lacks integrity may stop lying to a manager because she has been caught, but she still has low integrity and is likely to lie again when the threat of being caught is gone.
2. Trait-based systems, because they are vague, are more easily influenced by office politics, causing them to be less reliable as a source of information on an employee's true performance. The vagueness of these instruments allows managers to fill them out based on who they want to give a raise to, or feel should get a raise, rather than basing scores on specific behaviors that employees should or should not be engaging in. These systems are also more likely to leave a company open to discrimination claims because a manager can make biased decisions without having to back them up with specific behavioral information.

PROVIDE A MOTIVATING ENVIRONMENT

Maslow's hierarchy of needs is often depicted as a pyramid consisting of five levels: the four lower levels are grouped together as being associated with physiological needs, while the top level is termed *growth needs associated with psychological needs*. Deficiency needs must be met first. Once these are met, seeking to satisfy growth needs drives personal growth. The higher needs in this hierarchy only come into focus when the lower needs in the pyramid are satisfied. Once an individual has moved upward to the next level, needs in the lower level will no longer be prioritized. If a lower set of needs is no longer being met, the individual will temporarily reprioritize those needs by focusing attention on the unfulfilled needs, but will not permanently regress to the lower level. For instance, a businessman at the esteem level who is diagnosed with cancer will spend a great deal of time concentrating on his health (physiological needs), but will continue to value his work performance (esteem needs) and will likely return to work during periods of remission.

Deficiency Needs

The first four layers of the pyramid are what Maslow called "deficiency needs," or "D-needs": if they are not met, the body gives no indication of it physically, but the individual feels anxious and tense. The deficiency needs are survival needs, safety and security, love and belonging, and esteem.

Physiological Needs

These are the basic human needs for such things as food, warmth, water, and other bodily needs. If a person is hungry or thirsty or her body is chemically unbalanced, all of her energies turn toward remedying these deficiencies and other needs remain inactive. Maslow explains, "Anyone who attempts to make an emergency picture into a typical one and who will measure all of man's goals and desires by his [or her] behavior during extreme physiological deprivation, is certainly blind to many things. It is quite true that man [i.e., people] live(s) by bread alone—when there is no bread."

The physiological needs of the organism (those enabling homeostasis) take first precedence. These consist mainly of the following (in order of importance):

1. Breathing
2. Drinking
3. Eating
4. Excretion

If some needs are not fulfilled, a person's physiological needs take the highest priority. Physiological needs can control thoughts and behaviors and can cause people to feel sickness, pain, and discomfort.

Safety Needs

With their physical needs relatively satisfied, people's safety needs take over and dominate their behavior. These needs have to do with people's yearning for a predictable, orderly world in which injustice and inconsistency are under control, the familiar frequent, and the unfamiliar rare. In the world of work, these safety needs manifest themselves in such things as a preference for job security, grievance procedures for protecting the individual from unilateral authority, savings accounts, insurance policies, and the like.

For the most part, physiological and safety needs are reasonably well satisfied in the first world. The obvious exceptions, of course, are people outside the mainstream—the poor and the disadvantaged. If frustration has not led to apathy and weakness, such people still struggle to satisfy the basic physiological and safety needs. They are primarily concerned with survival: obtaining adequate food, clothing, and shelter, and seeking justice from the dominant societal groups.

Safety and security needs include the following:

1. Personal security from crime
2. Financial security
3. Health and well-being
4. Safety net against accidents and illness and the adverse impacts

Social Needs

After physiological and safety needs are fulfilled, the third layer of human needs is social. This psychological aspect of Maslow's hierarchy involves emotionally based relationships in general, such as the following:

1. Friendship
2. Intimacy
3. Having a supportive and communicative family

Humans need to feel a sense of belonging and acceptance, whether it comes from a large social group, such as clubs, office culture, religious groups, professional organizations, sports teams, or gangs ("safety in numbers"), or small social connections (family members, intimate partners, mentors, close colleagues, and confidants). They need to love and be loved (sexually and nonsexually) by others. In the absence of these elements, many people become susceptible to loneliness, social anxiety, and depression. This need for belonging can often overcome the physiological and security needs, depending on the strength of the peer pressure (e.g., an anorexic ignores the need to eat and the security of health for a feeling of control and belonging).

Esteem Needs

All humans have a need to be respected, to have self-esteem, to respect themselves, and to respect others. People need to engage themselves to gain recognition and have an activity or activities that give them a sense of contribution, and to feel accepted and self-valued, be it in a profession or hobby. Imbalances at this level can result in low self-esteem and inferiority complexes. People with low self-esteem need respect from others. They may seek fame or glory, which again depends on others. However, confidence, competence, and achievement only need one person, the self, and everyone else is inconsequential to one's own success. It may be noted, however, that many people with low self-esteem will not be able to improve their view of themselves simply by receiving fame, respect, and glory externally, but must first accept themselves internally. Psychological imbalances such as depression can also prevent one from obtaining self-esteem on both levels.

- *Growth needs*: Though the deficiency needs may be seen as "basic," and can be met and neutralized (i.e., they stop being motivators in one's life), self-actualization and transcendence are "being" or "growth needs" (also termed *B-needs*), that is, they are enduring motivations or drivers of behavior.
- *Cognitive needs*: Maslow believed that humans have the need to increase their intelligence and thereby chase knowledge. Cognitive needs are the expression of people's natural need to learn, explore, discover, and create to get a better understanding of the world around them.
- *Aesthetic needs*: Based on Maslow's beliefs, it is stated in the hierarchy that humans need beautiful imagery or something new and aesthetically pleasing to continue up toward self-actualization. Humans need to refresh themselves in the presence and beauty of nature while carefully absorbing and observing their surroundings to extract the beauty the world has to offer.

Self-Actualization

Self-actualization—a concept Maslow attributed to Kurt Goldstein, one of his mentors—is the instinctual need of humans to make the most of their abilities and to strive to be the best they can. It is working toward fulfilling our potential, toward becoming all that we are capable of becoming.

In Maslow's scheme, the final stage of psychological development comes when the individual feels assured that his physiological, security, affiliation and affection, self-respect, and recognition needs have been satisfied. As these become dormant, he becomes filled with a desire to realize all of his potential for being an effective, creative, mature human being. "What a man can be, he must be" is the way Maslow expresses it.

Maslow's need hierarchy is set forth as a general proposition and does not imply that everyone's needs follow the same rigid pattern. For some people, self-esteem seems to be a stronger motivation than love. Italian dictator Benito Mussolini, for example, alienated his closest friends by undertaking reckless military adventures to achieve status as a conqueror. (This example can also be used to illustrate the means-to-an-end dilemma of human motivation. That is, Mussolini may have reached for status as a means of gaining the affection of Adolf Hitler. More will be said about this problem later.) For some people, the need to create is often a stronger motivation than the need for food and safety. Thus, the artist living in poverty is a classic example of reversing the standard hierarchy of needs. Similarly, persons who have suffered hunger or some other deprivation for protracted periods may live happily for the rest of their lives if only they can get enough of what they lacked. In this case, the level of aspiration may have become permanently lowered and the higher-order, less proponent needs may never become active. There are also cases of people martyring themselves for causes and suffering all kinds of deprivations, particularly in the physiological, safety, and sometimes social categories, to achieve their goals.

Herzberg proposed the *motivation-hygiene theory*, also known as the *two-factor theory* (1959) of job satisfaction. According to his theory, people are influenced by two factors:

1. *Satisfaction,* is primarily the result of the *motivator factors.* These factors help increase satisfaction but have little effect on dissatisfaction.
   A. Motivator factors
      I. Achievement
      II. Recognition
      III. Work itself
      IV. Responsibility
      V. Promotion
      VI. Growth

2. *Dissatisfaction* is primarily the result of hygiene factors. These factors, if absent or inadequate, cause dissatisfaction, but their presence has little effect on long-term satisfaction.
   A. Hygiene factors
      I. Pay and benefits
      II. Company policy and administration
      III. Relationships with coworkers
      IV. Physical environment
      V. Supervision
      VI. Status
      VII. Job security

See Table 5.1 for a comparison of Maslow's and Herzberg's theories.

TABLE 5.1

Maslow and Herzberg Comparison

Maslow Need Hierarchy Theory		Herzberg Motivation-Hygiene Theory
Self-actualization	*Motivational*	Work itself
– Self-fulfillment		Achievement
– Challenge		Growth
Esteem (ego)		Advancement
– Recognition		Recognition
– Confidence		Status
Social	*Maintenance*	Interpersonal relations
– Acceptance		– Supervisor
– Belonging		– Subordinates
– Affection		– Peers
– Participation		Supervision
Safety		Company policy and administration
– Security		Job security
– Protection		Working conditions
Physiological		Salary
– Food, water, and sleep		Personal life

SUMMARY

The motivation process begins with proper leadership. Leadership should lead to an appropriate motivating environment which provides consistency in purpose and sincerity, and where goals are realistic and there are no hidden agendas. Management needs to be aware that each individual has different needs depending on her stage in life and should be treated accordingly. Operator dissatisfaction will be reflected in her work when these needs are not recognized.

When management communicates in only one direction to employees, mostly through threats and very few rewards, employees will form informal objectives that are diametrically opposed to those of the organization, resulting in ever-decreasing levels of output value.

REVIEW QUESTIONS

1. What is the purpose of management?
2. Describe leadership.
3. Define different types of leadership and management styles.
4. Define performance appraisal and its purpose.
5. Describe the different levels of Maslow's hierarchy.
6. Describe Herzberg's motivation-hygiene theory.
7. Describe the differences between Maslow and Herzberg.

6

Special Topics in Quality

OVERVIEW OF STATISTICAL METHODS

Statistics is a mathematical science pertaining to the collection, analysis, interpretation or explanation, and presentation of data. It is applicable to a wide variety of academic disciplines, from the natural and social sciences to the humanities, and to government and business.

Statistical methods can be used to summarize or describe a collection of data; this is called *descriptive statistics*. In addition, patterns in the data may be modeled in a way that accounts for randomness and uncertainty in the observations, and then used to draw inferences about the process or population being studied; this is called *inferential statistics*. Both descriptive and inferential statistics comprise applied statistics. There is also a discipline called *mathematical statistics*, which is concerned with the theoretical basis of the subject.

The word *statistics* is also the plural of *statistic* (singular), which refers to the result of applying a statistical algorithm to a set of data, as in economic statistics, crime statistics, and so on.

History

Some scholars pinpoint the origin of statistics to 1662, with the publication of "Observations on the Bills of Mortality" by John Graunt. Early applications of statistical thinking revolved around the needs of states to base policy on demographic and economic data. The scope of the discipline of statistics broadened in the early nineteenth century to include the collection and analysis of data in general. Today, statistics is widely employed in government, business, and the natural and social sciences.

Because of its empirical roots and its applications, statistics is generally considered not to be a subfield of pure mathematics, but rather a distinct branch of applied mathematics. Its mathematical foundations were laid in the seventeenth century with the development of probability theory by Blaise Pascal and Pierre de Fermat. Probability theory arose from the study of games of chance. The method of least squares was first described by Carl Friedrich Gauss around 1794. The use of modern computers has expedited large-scale statistical computation, and has also made possible new methods that are impractical to perform manually.

Overview

In applying statistics to a scientific, industrial, or societal problem, one begins with a process or population to be studied. This might be a population of people in a country, of crystal grains in a rock, or of goods manufactured by a particular factory during a given period. It may instead be a process observed at various times; data collected about this kind of "population" constitute what is called a *time series.*

For practical reasons, rather than compiling data about an entire population, one usually studies a chosen subset of the population, called a *sample.* Data are collected about the sample in an observational or experimental setting. The data are then subjected to statistical analysis, which serves two related purposes: description and inference.

Descriptive statistics can be used to summarize the data, either numerically or graphically, to describe the sample. Basic examples of numerical descriptors include the mean and standard deviation. Graphical summarizations include various kinds of charts and graphs.

Inferential statistics is used to model patterns in the data, accounting for randomness and drawing inferences about the larger population. These inferences may take the form of answers to yes/no questions (hypothesis testing), estimates of numerical characteristics (estimation), descriptions of association (correlation), or modeling of relationships (regression). Other modeling techniques include analysis of variance (ANOVA), time series, and data mining.

The concept of correlation is particularly noteworthy. Statistical analysis of a data set may reveal that two variables (that is, two properties of the population under consideration) tend to vary together, as if they are connected. For example, a study of annual income and age of death among people might find that poor people tend to have shorter lives than affluent people. The two variables are said to be correlated (which is a positive correlation in this case). However, one cannot immediately infer the existence of a causal relationship between the two variables. The correlated phenomena could be caused by a third, previously unconsidered phenomenon, called a *lurking variable* or *confounding variable*.

If the sample is representative of the population, then inferences and conclusions made from the sample can be extended to the population as a whole. A major problem lies in determining the extent to which the chosen sample is representative. Statistics offers methods to estimate and correct for randomness in the sample and in the data collection procedure, as well as methods for designing robust experiments in the first place. The fundamental mathematical concept employed in understanding such randomness is probability. Mathematical statistics (also called *statistical theory*) is the branch of applied mathematics that uses probability theory and analysis to examine the theoretical basis of statistics.

The use of any statistical method is valid only when the system or population under consideration satisfies the basic mathematical assumptions of the method. Misuse of statistics can produce subtle but serious errors in description and interpretation—subtle in the sense that even experienced professionals sometimes make such errors, and serious in the sense that they may affect, for instance, social policy, medical practice, and the reliability of structures such as bridges. Even when statistics is correctly applied, the results can be difficult for the nonexpert to interpret. For example, the statistical significance of a trend in the data, which measures the extent to which the trend could be caused by random variation in the sample, may not agree with one's intuitive sense of its significance. The set of basic statistical skills (and skepticism) needed by people to deal with information in their everyday lives is referred to as *statistical literacy*.

Statistical Methods

Experimental and Observational Studies

A common goal for a statistical research project is to investigate causality, and in particular to draw a conclusion on the effect of changes in the values of predictors or independent variables on response or dependent variables. There are two major types of causal statistical studies, experimental studies and observational studies. In both types of studies, the effects of differences of an independent variable (or variables) on the behavior of the dependent variable are observed. The difference between the two types lies in how the study is actually conducted. Each can be very effective.

An experimental study involves taking measurements of the system under study, manipulating the system, and then taking additional measurements using the same procedure to determine if the manipulation has modified the values of the measurements. In contrast, an observational study does not involve experimental manipulation. Instead, data are gathered and correlations between predictors and response are investigated.

An example of an experimental study is the famous Hawthorne studies, which attempted to test the changes to the working environment at the Hawthorne plant of the Western Electric Company. The researchers were interested in determining whether increased illumination would increase the productivity of the assembly line workers. The researchers first measured the productivity in the plant, then modified the illumination in an area of the plant and checked if the changes in illumination affected the productivity. It turned out that the productivity indeed improved (under the experimental conditions). However, the study is heavily criticized today for errors in experimental procedures, specifically for the lack of a control group and blindedness.

An example of an observational study is a study which explores the correlation between smoking and lung cancer. This type of study typically uses a survey to collect observations about the area of interest, and then performs statistical analysis of the observational data. In this case, the researchers would collect observations of both smokers and nonsmokers, perhaps through a case-control study, and then look for the number of cases of lung cancer in each group.

The basic steps of an experiment are as follows:

- Planning the research, including determining information sources, research subject selection, and ethical considerations for the proposed research and method
- Designing experiments, concentrating on the system model and the interaction of independent and dependent variables

- Summarizing a collection of observations to feature their commonality by suppressing details (descriptive statistics)
- Reaching consensus about what the observations tell about the world being observed (statistical inference)
- Documenting and presenting the results of the study

Levels of Measurement

There are four types of measurements, levels of measurement, or measurement scales used in statistics: nominal, ordinal, interval, and ratio. They have different degrees of usefulness in statistical research. Ratio measurements have both a zero value and the distances between different measurements defined; they provide the greatest flexibility in statistical methods that can be used for analyzing the data. Interval measurements have meaningful distances between measurements defined, but have no meaningful zero value (as in the case with IQ measurements or with temperature measurements in Fahrenheit). Ordinal measurements have imprecise differences between consecutive values, but have a meaningful order to those values. Nominal measurements have no meaningful rank order among values.

Since variables conforming only to nominal or ordinal measurements cannot be reasonably measured numerically, sometimes they are referred to as categorical variables, whereas ratio and interval measurements are grouped together as quantitative or continuous variables due to their numerical nature.

Statistical process control (SPC) is an effective method of monitoring a process through the use of control charts. Control charts enable the use of objective criteria for distinguishing background variation from events of significance based on statistical techniques. Much of SPC's power lies in the ability to monitor both the process center and its variation about that center. By collecting data from samples at various points within the process, variations in the process that may affect the quality of the end product or service can be detected and corrected, thus reducing waste as well as the likelihood that problems will be passed on to the customer. With its emphasis on early detection and prevention of problems, SPC has a distinct advantage over quality methods such as inspection, that apply resources to detecting and correcting problems in the end product or service.

In addition to reducing waste, SPC can lead to a reduction in the time required to produce the product or provide the service from end to end.

TABLE 6.1

Statistical Formulas

Statistic	Formula	Used for
Continuous Statistics		
Mean	$\mu = \dfrac{\sum x}{n}$	The center of a set of data
Range	$r = x_{\max} - x_{\min}$	The dispersion of data around the center
Variance	$\sigma^2 = \dfrac{\sum(\mu - x)^2}{n-1}$	The dispersion of data around the center
Standard Deviation	$\sigma = \sqrt{\dfrac{\sum(\mu - n)^2}{n-1}}$	The dispersion of data around the center
Normal Distribution	$f(x) = \dfrac{1}{\sigma\sqrt{2\pi}} \bullet e^{-\frac{1}{2}\left(\frac{\mu - x}{\sigma}\right)^2}$	Used to perform estimations
Standard Normal Value	$z = \dfrac{\mu - x}{\sigma}$	Used to determine normalcy
Hypothesis Test of Means	$z = \dfrac{\mu_1 - \mu_2}{\sigma/\sqrt{n}}$	Used to determine differences

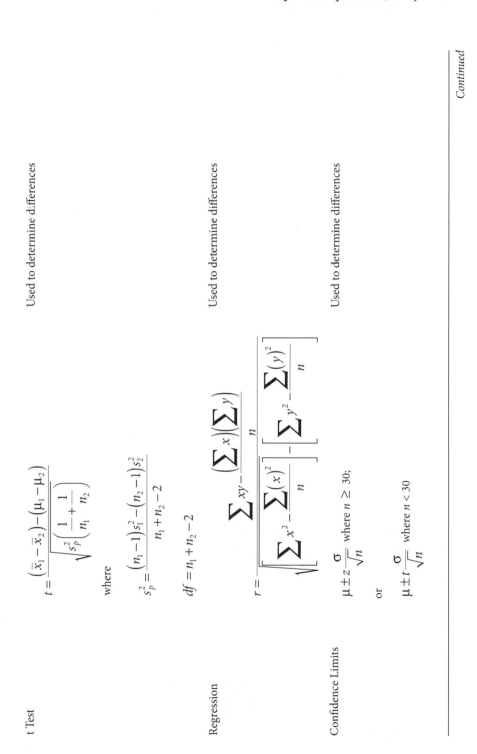

t Test

$$t = \frac{(\bar{x}_1 - \bar{x}_2) - (\mu_1 - \mu_2)}{\sqrt{s_p^2\left(\dfrac{1}{n_1} + \dfrac{1}{n_2}\right)}}$$

where

$$s_p^2 = \frac{(n_1-1)s_1^2 - (n_2-1)s_2^2}{n_1 + n_2 - 2}$$

$$df = n_1 + n_2 - 2$$

Used to determine differences

Regression

$$r = \frac{\displaystyle\sum xy - \frac{\left(\sum x\right)\left(\sum y\right)}{n}}{\sqrt{\left[\displaystyle\sum x^2 - \frac{\left(\sum x\right)^2}{n}\right]\left[\displaystyle\sum y^2 - \frac{\left(\sum y\right)^2}{n}\right]}}$$

Used to determine differences

Confidence Limits

$$\mu \pm z \frac{\sigma}{\sqrt{n}} \text{ where } n \geq 30;$$

or

$$\mu \pm t \frac{\sigma}{\sqrt{n}} \text{ where } n < 30$$

Used to determine differences

Continued

TABLE 6.1. (continued)
Statistical Formulas

Statistic	Formula	Used for
Discrete Statistics		
Proportion	$p = \sum_{x}^{r}$	Used to determine percentage nonconforming
Binomial Distribution	$\mu = \dfrac{n!}{r!(n-r)!} p^r q^{(n-r)}$	Used to determine average percentage nonconforming
Poisson Distribution	$\mu = \dfrac{\lambda^x e^{-\lambda}}{x!}$	Used to determine average percentage nonconforming
Hypergeometric Distribution	$\mu = \dfrac{C_d^D C_{n-d}^{N-D}}{C_n^N}$ where n = sample size, N = lot size, D = number of failures, and d = probability of a failure	Used in statistical sampling
$p = p_0$ Hypothesis Test	$\mu = \dfrac{\hat{p} - p_0}{\sqrt{\dfrac{p_0(1-p_0)}{n}}}$	Used to determine differences

$p_1 - p_2 = 0$
Hypothesis Test

$$\mu = \frac{(\hat{p}_1 - \hat{p}_2) - 0}{\sqrt{\hat{p}(1-\hat{p})\left(\dfrac{1}{n_1} + \dfrac{1}{n_2}\right)}}$$

where $n \times \hat{p}$ and $n(1-\hat{p})$ are at least 5

Used to determine differences

Chi-Square Distribution

$$x^2 = \sum \frac{(O_i - E_i)^2}{E_i}$$

Where $E_i = \dfrac{R_1 \bullet C_i}{n}$

Used to analyze survey data

Confidence
Limits

$$\mu = z\sqrt{\frac{\hat{p}(1-\hat{p})}{n}}$$

where $n\hat{p}$ and $n(1-\hat{p})$ are at least 5

Used to determine differences

This is partially due to a diminished likelihood that the final product will have to be reworked, but it may also result from using SPC data to identify bottlenecks, wait times, and other sources of delays within the process. Process cycle time reductions coupled with improvements in yield have made SPC a valuable tool from both a cost reduction and a customer satisfaction standpoint.

History of SPC

Statistical process control was pioneered by Walter A. Shewhart in the early 1920s. W. Edwards Deming later applied SPC methods in the United States during World War II, thereby successfully improving quality in the manufacture of munitions and other strategically important products. Deming was also instrumental in introducing SPC methods to Japanese industry after the war had ended.

Shewhart created the basis for the control chart and the concept of a state of statistical control by carefully designed experiments. While Dr. Shewhart drew from pure mathematical statistical theories, he understood that data from physical processes seldom produce a *normal distribution curve* (a Gaussian distribution, also commonly referred to as a *bell curve*). He discovered that observed variation in manufacturing data did not always behave the same way as with data in nature (for example, Brownian motion of particles). Dr. Shewhart concluded that while every process displays variation, some processes display controlled variation that is natural to the process (common causes of variation), while others display uncontrolled variation that is not present in the process causal system at all times (special causes of variation).

General

The following description relates to manufacturing rather than to the service industry, although the principles of SPC can be successfully applied to either. SPC has also been successfully applied to detecting changes in organizational behavior, with social network change detection introduced by McCulloh (2007).

In mass manufacturing, the quality of the finished article was traditionally achieved through postmanufacturing inspection of the product, accepting or rejecting each article (or samples from a production lot) based on how well it met its design specifications. In contrast, SPC uses statistical tools to observe the performance of the production process in order to predict significant deviations that may later result in rejected product.

Two kinds of variations occur in all manufacturing processes; both these process variations cause subsequent variations in the final product: The first are known as *natural* or *common causes* of variation and may be variations in temperature, specifications of raw materials or electrical current, and so on. These variations are small, and are generally near to the average value. The pattern of variation will be similar to those found in nature, and the distribution forms the bell-shaped normal distribution curve (see Figure 6.1). The second kind is known as *special causes*, and happens less frequently than the first.

For example, a breakfast cereal–packaging line may be designed to fill each cereal box with 500 grams of product, but some boxes will have slightly more than 500 grams, and some will have slightly less, in accordance with a distribution of net weights. If the production process, its inputs, or its environment changes (for example, the machines doing the manufacture begin to wear), this distribution can change. For example, as its cams and pulleys wear out, the cereal-filling machine may start putting more cereal into each box than specified. If this change is allowed to continue unchecked, more and more product will be produced that falls outside the tolerances of the manufacturer or consumer, resulting in waste. While in this case, the waste is in the form of "free" product for the consumer, typically waste consists of rework or scrap.

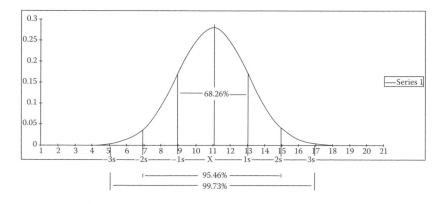

FIGURE 6.1
Normal curve.

TABLE 6.2

Statistical Methods Applied to Operations

Statistic	Formula	Used For
Attribute Control Charts		
P Chart	$CL = \bar{p} \pm 3\sqrt{\dfrac{\bar{p}(1-\bar{p})}{n}}$	Process; tracking proportion nonconforming
C Chart	$CL = \bar{c} \pm 3\sqrt{\bar{c}}$	Process: tracking nonconformities
Np Chart	$CL = n\bar{p} \pm 3\sqrt{n\bar{p}(1-p)}$	Process: tracking multiple nonconformities per unit
U Chart	$CL = \bar{u} \pm 3\sqrt{\dfrac{\bar{u}}{n}}$	Process: tracking multiple nonconformities per sample
Variable Control Charts		
$\bar{X}$ Chart	$CL = \bar{\bar{X}} \pm A_2 \bar{R}$	Product: tracking product consistency and accuracy
$\bar{R}$ Chart	$CL = D_4 \bar{R}$	Product: tracking product consistency and accuracy
$\bar{\bar{X}}_s$ Chart	$CL = \bar{\bar{X}} \pm A_3 \bar{S}$	Product: tracking product consistency and accuracy
$\bar{S}$ Chart	$CL = B_4 \bar{S}$	Product: tracking product consistency and accuracy

Where n = sample size, $\bar{S}$ = average of sample standard deviations, A_2, B_4, D_4 = constants, $\bar{\bar{X}}$ = average of samples averages, and $\bar{R}$ = average of sample ranges.

TABLE 6.3

SPC Constants

N	D4	B4	A2	k1	k2	d2
2	3.268	3.267	1.860	4.56	3.65	1.128
3	2.574	2.568	1.023	3.05	2.70	1.693
4	2.282	2.266	0.729	2.50	2.30	2.059
5	2.115	2.089	0.577			2.326
6	2.004	1.970	0.483			2.534

By observing at the right time what happened in the process that led to a change, the quality engineer or any member of the team responsible for the production line can troubleshoot the root cause of the variation that has crept into the process and correct the problem.

TABLE 6.4.

Quality-Engineering Formulas

Name	Equation	Used For
Process Capability	$Cp = \dfrac{US - LS}{6\sigma}$	Product design versus process
Upper Capability	$Cp_u = \dfrac{US - \overline{\overline{X}}}{3\hat{\sigma}}$	Product design versus process
Lower Capability	$Cp_L = \dfrac{\overline{\overline{X}} - LS}{3\hat{\sigma}}$	Product design versus process
Sigma Estimation	$\hat{\sigma} = \overline{R}\big/ d_2$	Product design versus process
Part per Million	$PPM = \dfrac{R_e}{n} \times 1{,}000{,}000$	Process
Repeatability and Reproducibility	$R\&R = \sqrt{\left(\overline{\overline{R}} \times k_1\right)^2 + \left(\left(\overline{X}_{diff} \times k_2\right)^2 - \left[\dfrac{\left(\overline{R} \times k_1\right)^2}{(n \times r)}\right]\right)}$ $\%R\&R = 100\left(\dfrac{(R\&R)}{(US - LS)}\right)$ $\sigma_c = \sqrt{\left(\sigma_a\right)^2 + \left(\sigma_b\right)^2}$	Measurement analysis

Where US = upper specification, LS = lower specification, n = number of parts, r = number of trials, R_e = number of nonconformities, and k_1, k_2, and d_2 = constants.

TABLE 6.5

Statistical Sampling Plan

Lot Size		Acceptable Nonconformance Levels and Sample Sizes			
From	**To**	**2.5%**	**4.0%**	**6.5%**	**10%**
2	8	5	3	2	2
9	15	5	3	2	2
16	25	5	3	3	2
26	50	5	5	5	3
51	90	7	6	5	4
91	150	11	7	6	5
151	280	13	10	7	6
281	500	16	11	9	7
501	1,200	19	15	11	8
1,201	3,200	23	18	13	9
3,201	10,000	29	22	15	9
10,001	35,000	35	29	15	9
35,001	150,000	40	29	15	9
150,001	500,000	40	29	15	9
500,001	>500,001	40	29	15	9

Reject on one nonconformity and accept on zero nonconformities.

SPC indicates when an action should be taken in a process, but it also indicates when *no* action should be taken. An example is a person who would like to maintain a constant body weight and takes weight measurements weekly. A person who does not understand SPC concepts might start dieting every time his or her weight increased, or eat more every time his or her weight decreased. This type of action could be harmful and possibly generate even more variation in body weight. SPC would account for normal weight variation and better indicate when the person is in fact gaining or losing weight.

RISK ANALYSIS

Risk analysis is the science of risks and their probability and evaluation.

The term *cindynics* (from the Greek *kindunos*, "danger") has been proposed for this field. This term is used in France, but has not been widely adopted in the English-speaking world. Probabilistic risk assessment is the analysis strategy usually employed in science and engineering.

Risk Analysis and the Risk Workshop

As part of the risk management process, risk analysis for each project should be performed. The data from this would be based on risk discussion workshops to identify potential issues and risks ahead of time before these were to pose negative cost and/or schedule impacts (see the article on cost contingency for a discussion of the estimation of cost impacts).

The risk workshops should be chaired by a small group, ideally between 6 and 10 individuals from the various departmental functions (e.g., project manager, construction manager, site superintendent, and representatives from operations, procurement, [project] controls, etc.) so as to cover every risk element from different perspectives.

The outcome of the risk analysis would be the creation and review of the risk register to identify and quantify risk elements to the project and their potential impact.

Given that risk management is a continuous and iterative process, the risk workshop members would regroup at regular intervals and project milestones to review the risk register mitigation plans, make changes to it as appropriate, and, following those changes, rerun the risk model. By constantly monitoring risks, they can successfully mitigate them, resulting in a cost and schedule savings with a positive impact on the project.

Probabilistic risk assessment (PRA) (or probabilistic safety assessment or analysis) is a systematic and comprehensive methodology to evaluate risks associated with a complex engineered technological entity (such as airliners or nuclear power plants). Risk in a PRA is defined as a feasible detrimental outcome of an activity or action. In a PRA, risk is characterized by two quantities:

The magnitude (severity) of the possible adverse consequence(s)
The likelihood (probability) of occurrence of each consequence

Consequences are expressed numerically (e.g., the number of people potentially hurt or killed), and their likelihoods of occurrence are expressed as probabilities or frequencies (i.e., the number of occurrences or the probability of occurrence per unit time). The total risk is the sum of the products of the consequences multiplied by their probabilities. The spectrums of risks across classes of events are also of concern, and are usually controlled in licensing processes (it would be of concern if rare but high-consequence events were found to dominate the overall risk).

Probabilistic risk assessment usually answers three basic questions:

> What can go wrong with the studied technological entity, or what are the initiators or initiating events (undesirable starting events) that lead to adverse consequence(s)?
> What and how severe are the potential detriments or the adverse consequences that the technological entity may be eventually subjected to as a result of the occurrence of the initiator?
> How likely to occur are these undesirable consequences, or what are their probabilities or frequencies?

Two common methods of answering this last question are event tree analysis and fault tree analysis—for explanations of these, see safety engineering.

In addition to the above methods, PRA studies require special but often very important analysis tools like human reliability analysis (HRA) and common cause or failure (CCF) analysis. HRA deals with methods for modeling human error, while CCF analysis deals with methods for evaluating the effect of intersystem and intrasystem dependencies which tend to cause simultaneous failures and thus significant increases in overall risk.

PRA studies have been successfully performed for complex technological systems at all phases of the life cycle from concept definition and predesign through safe removal from operation. For example, the Nuclear Regulatory Commission (NRC) required that each nuclear power plant in the United States perform an individual plant examination (IPE) to identify and quantify plant vulnerabilities to hardware failures and human faults in design and operation. Although no method was specified for performing such an evaluation, the NRC requires risk analysis (Business).

Risk analysis is a technique to identify and assess factors that may jeopardize the success of a project or the achievement of a goal. This technique also helps to define preventive measures to reduce the probability of these factors from occurring, as well as identify countermeasures to successfully deal with these constraints when they develop to avert possible negative effects on the competitiveness of the company.

One of the more popular methods to perform a risk analysis in the computer field is called the *facilitated risk analysis process* (FRAP).

Facilitated Risk Analysis Process

FRAP analyzes one system, application, or segment of business processes at a time.

Practitioners of FRAP believe that additional efforts to develop precisely quantified risks are not cost-effective because

such estimates are time consuming;

risk documentation becomes too voluminous for practical use; and

specific loss estimates are generally not needed to determine if controls are needed.

After identifying and categorizing risks, the FRAP team identifies the controls that could mitigate the risk. The decision for what controls are needed lies with the business manager. The team's conclusions as to what risks exist and what controls are needed are documented, along with a related action plan for control implementation.

Three of the most important risks a software company faces are unexpected changes in (1) revenue, (2) costs from those budgeted, and (3) the amount of specialization of the software planned. Risks that affect revenues can be unanticipated competition, privacy, intellectual property rights problems, and unit sales that are less than forecasted; unexpected development costs also create risk that can be in the form of more rework than anticipated, security holes, and privacy invasions.

Narrow specialization of software with a large amount of research and development expenditures can lead to both business and technological risks, since specialization does not lead to lower unit costs of software (Messerschmidt and Szyperski 2004). Combined with the decrease in the potential customer base, specialization risk can be significant for a software firm. After probabilities of scenarios have been calculated with risk analysis, the process of risk management can be applied to help manage the risk.

Methods like applied information economics add to and improve on risk analysis methods by introducing procedures to adjust subjective probabilities, compute the value of additional information, and use the results in part of a larger portfolio management problem.

RELIABILITY ENGINEERING

Reliability Theory

Reliability theory is the foundation of reliability engineering. For engineering purposes, *reliability* is defined as the probability that a device will perform its intended function during a specified period of time under stated conditions.

Mathematically, this may be expressed as

$$R_{(t)} = \int_t^{\infty} f(x)dx$$

where $f(x)$ is the failure probability density function, and t is the length of the period (which is assumed to start from time zero).

Reliability engineering is concerned with four key elements of this definition:

First, reliability is a probability. This means that failure is regarded as a random phenomenon: it is a recurring event, and we do not express any information on individual failures, the causes of failures, or relationships between failures, except that the likelihood for failures to occur varies over time according to the given probability function. Reliability engineering is concerned with meeting the specified probability of success, at a specified statistical confidence level.

Second, reliability is predicated on "intended function": generally, this is taken to mean operation without failure. However, even if no individual part of the system fails, but the system as a whole does not do what was intended, then the failure is still charged against the system reliability. The system requirements specification is the criterion against which reliability is measured.

Third, reliability applies to a specified period of time. In practical terms, this means that a system has a specified chance that it will operate without failure before a specified time. Reliability engineering ensures that components and materials will meet the requirements during the specified time. Units other than time may sometimes be used: the automotive industry might specify reliability in terms of miles; the military might specify reliability of a gun for a certain number of rounds fired; or a piece of mechanical equipment may have a reliability rating value in terms of cycles of use.

Fourth, reliability is restricted to operation under stated conditions. This constraint is necessary because it is impossible to design a system for unlimited conditions. A Mars Rover will have different specified conditions than the family car. The operating environment must be addressed during design and testing.

Reliability Program Plan

Many tasks, methods, and tools can be used to achieve reliability. Every system requires a different level of reliability. A commercial airliner must operate under a wide range of conditions. The consequences of failure are grave, but there is a correspondingly higher budget. A pencil sharpener may be more reliable than an airliner, but it has a much different set of operational conditions, insignificant consequences of failure, and a much lower budget.

A reliability program plan is used to document exactly what tasks, methods, tools, analyses, and tests are required for a particular system. For complex systems, the reliability program plan is a separate document. For simple systems, it may be combined with the systems engineering management plan. The reliability program plan is essential for a successful reliability program and is developed early during system development. It specifies not only what the reliability engineer does, but also the tasks performed by others. The reliability program plan is approved by top program management.

Reliability Requirements

For any system, one of the first tasks of reliability engineering is to adequately specify the reliability requirements. Reliability requirements address the system itself, test and assessment requirements, and associated tasks and documentation. Reliability requirements are included in the appropriate system and subsystem requirement specifications, test plans, and contract statements.

System Reliability Parameters

Requirements are specified using reliability parameters. The most common reliability parameter is mean time between failure (MTBF), which can also be specified as the failure rate or the number of failures during a given period. These parameters are very useful for systems that are operated on a regular basis, such as most vehicles, machinery, and electronic equipment. Reliability increases as the MTBF increases. The MTBF is usually specified in hours, but can also be used with other units of measurement such as miles or cycles.

In other cases, reliability is specified as the probability of mission success. For example, reliability of a scheduled aircraft flight can be specified as a dimensionless probability or a percentage.

A special case of mission success is the single-shot device or system. These are devices or systems that remain relatively dormant and operate only once. Examples include automobile airbags, thermal batteries, and missiles. Single-shot reliability is specified as a probability of success, or is subsumed into a related parameter. Single-shot missile reliability may be incorporated into a requirement for the probability of hit.

For such systems, the probability of failure on demand (PFD) is the reliability measure. This PFD is derived from failure rate and mission time for nonrepairable systems. For repairable systems, it is obtained from failure rate and mean time to recovery (MTTR) and test interval. This measure may not be unique for a given system, as the measure depends on the kind of demand. In addition to system-level requirements, reliability requirements may be specified for critical subsystems. In all cases, reliability parameters are specified with appropriate statistical confidence intervals.

Reliability Modeling

Reliability modeling is the process of predicting or understanding the reliability of a component or system. Two separate fields of investigation are common: the physics of failure approach uses an understanding of the failure mechanisms involved, such as crack propagation or chemical corrosion; and the parts stress modeling approach is an empirical method for prediction based on counting the number and type of components of the system, and the stress they undergo during operation.

For systems with a clearly defined failure time (which is sometimes not given for systems with a drifting parameter), the empirical distribution function of these failure times can be determined. This is done in general in an accelerated experiment with increased stress. These experiments can be divided into two main categories.

Early failure rate studies determine the distribution with a decreasing failure rate over the first part of the bathtub curve. Here, in general, only moderate stress is necessary. The stress is applied for a limited period of time in what is called a *censored test*. Therefore, only the part of the distribution with early failures can be determined.

In so-called zero-defect experiments, only limited information about the failure distribution is acquired. Here the stress, stress time, or the sample size is so low that not a single failure occurs. Due to the insufficient sample size, only an upper limit of the early failure rate can be determined. At any rate, it looks good for the customer if there are no failures.

In a study of the intrinsic failure distribution, which is often a material property, higher stresses are necessary to achieve failure in a reasonable period of time. Several degrees of stress have to be applied to determine an acceleration model. The empirical failure distribution is often parameterized with a Weibull or a log-normal model.

It is a general praxis to model the early failure rate with an exponential distribution. This less complex model for the failure distribution has only one parameter: the constant failure rate. In such cases, the chi-square distribution can be used to find the goodness of fit for the estimated failure rate. Compared to a model with a decreasing failure rate, this is quite pessimistic. Combined with a zero-defect experiment, this becomes even more pessimistic. The effort is greatly reduced in this case: one does not have to determine a second model parameter (e.g., the shape parameter of a Weibull distribution) or its confidence interval (e.g., by a maximum likelihood approach, or MLE), and the sample size is much smaller.

Reliability Test Requirements

Because reliability is a probability, even highly reliable systems have some chance of failure. However, testing reliability requirements is problematic for several reasons. A single test is insufficient to generate enough statistical data. Multiple tests or long-duration tests are usually very expensive. Some tests are simply impractical. Reliability engineering is used to design a realistic and affordable test program that provides enough evidence that the system meets its requirements. Statistical confidence levels are used to address some of these concerns. A certain parameter is expressed along with a corresponding confidence level: for example, an MTBF of 1,000 hours at a 90 percent confidence level. From this specification, the reliability engineer can design a test with explicit criteria for the number of hours and number of failures until the requirement is met or failed.

Actual mean time between failures is calculated as follows:

$$MTBF = \frac{\sum units \times \sum hours}{\sum Failures} \quad \text{(60\% failure rate)}$$

System survivability is calculated as follows:

$$R_s = e^{\left(-t/MTBF\right)}$$

where $-t$ is the operational hours of concern and $e = 2.18$.

The combination of reliability parameter value and confidence level greatly affects the development cost and the risk to both the customer and producer. Care is needed to select the best combination of requirements. Reliability testing may be performed at various levels, such as the component, subsystem, and system levels. Also, many factors must be addressed during testing, such as extreme temperature and humidity, shock, vibration, and heat. Reliability engineering determines an effective test strategy so that all parts are exercised in relevant environments. For systems that must last many years, reliability engineering may be used to design an accelerated life test.

Requirements for Reliability Tasks

Reliability engineering must also address requirements for various reliability tasks and documentation during system development, testing, production, and operation. These requirements are generally specified in the contract statement of work and depend on how much leeway the customer wishes to provide to the contractor. Reliability tasks include various analyses, planning, and failure reporting. Task selection depends on the criticality of the system as well as cost. A critical system may require a formal failure-reporting and failure review process throughout development, whereas a noncritical system may rely on final test reports. The most common reliability program tasks are documented in reliability program standards, such as MIL-STD-785 (U.S. Air Force 1986) and IEEE 1332 (Institute of Electrical and Electronics Engineers 1998).

Design for Reliability

Design for reliability (DFR) is an emerging discipline that refers to the process of designing reliability into products. This process encompasses several tools and practices, and describes the order of their deployment that an organization needs to have in place in order to drive reliability into their products. Typically, the first step in the DFR process is to set the system's reliability requirements. Reliability must be "designed into" the system. During system design, the top-level reliability requirements are then allocated to subsystems by design and reliability engineers working together.

Reliability design begins with the development of a model. Reliability models use block diagrams and fault trees to provide a graphical means of evaluating the relationships between different parts of the system. These models incorporate predictions based on parts-count failure rates taken from historical data. While the predictions are often not accurate in an absolute sense, they are valuable to assess relative differences in design alternatives.

Fault Tree Diagrams

One of the most important design techniques is redundancy. This means that if one part of the system fails, there is an alternate success path, such as a backup system. An automobile brake light might use two light bulbs. If one bulb fails, the brake light still operates using the other bulb. Redundancy significantly increases system reliability, and is often the only viable means of doing so. However, redundancy is difficult and expensive, and is therefore limited to critical parts of the system. Another design technique, the physics of failure, relies on understanding the physical processes of stress, strength, and failure at a very detailed level. The material or component can then be redesigned to reduce the probability of failure. Another common design technique is component derating: selecting components whose tolerance significantly exceeds the expected stress, as by using a heavier gauge wire that exceeds the normal specification for the expected electrical current.

Many tasks, techniques, and analyses are specific to particular industries and applications. Commonly these include the following:

Built-in test (BIT)
Failure mode and effects analysis (FMEA)
Reliability simulation modeling
Thermal analysis
Reliability block diagram analysis
Fault tree analysis
Sneak circuit analysis
Accelerated testing
Reliability growth analysis
Weibull analysis
Electromagnetic analysis
Statistical interference

Results are presented during the system design reviews and logistics reviews. Reliability is just one requirement among many system requirements. Engineering trade studies are used to determine the optimum balance between reliability and other requirements and constraints.

Reliability Testing

A Reliability Sequential Test Plan

The purpose of reliability testing is to discover potential problems with the design as early as possible and, ultimately, provide confidence that the system meets its reliability requirements.

Reliability testing may be performed at several levels. Complex systems may be tested at the component, circuit board, unit, assembly, subsystem, and system levels. (The test-level nomenclature varies among applications.) For example, performing environmental stress–screening tests at lower levels, such as with piece parts or small assemblies, catches problems before they cause failures at higher levels. Testing proceeds during each level of integration through full-up system testing, developmental testing, and operational testing, thereby reducing program risk. System reliability is calculated at each test level. Reliability growth techniques and failure-reporting, analysis, and corrective action systems (FRACAS) are often employed to improve reliability as testing progresses. The drawbacks to such extensive testing are time and expense. Customers may choose to accept more risk by eliminating some or all lower levels of testing.

It is not always feasible to test all system requirements. Some systems are prohibitively expensive to test; some failure modes may take years to observe; some complex interactions result in a huge number of possible test cases; and some tests require the use of limited test ranges or other resources. In such cases, different approaches to testing can be used, such as accelerated life testing, the design of experiments, and simulations.

The desired level of statistical confidence also plays an important role in reliability testing. Statistical confidence is increased by increasing either the test time or the number of items tested. Reliability test plans are designed to achieve the specified reliability at the specified confidence level with the minimum number of test units and test time. Different test plans result in different levels of risk to the producer and consumer. The desired reliability, statistical confidence, and risk levels for each side influence the ultimate test plan. Good test requirements ensure that the customer and developer agree in advance on how reliability requirements will be tested.

A key aspect of reliability testing is to define *failure*. Although this may seem obvious, there are many situations where it is not clear whether a failure is really the fault of the system. Variations in test conditions, operator differences, weather, and unexpected situations create differences between the customer and the system developer. One strategy to address this issue is to use a scoring conference process. A scoring conference includes representatives from the customer, the developer, the test organization, and the reliability organization, and sometimes independent observers. The scoring conference process is defined in the statement of work. Each test case is considered by the group and "scored" as a success or failure. This scoring is the official result used by the reliability engineer.

As part of the requirements phase, the reliability engineer develops a test strategy with the customer. The test strategy makes trade-offs between the needs of the reliability organization, which wants as much data as possible, and constraints such as cost, schedule, and available resources. Test plans and procedures are developed for each reliability test, and results are documented in official reports.

Accelerated Testing

The purpose of accelerated life testing is to induce field failure in the laboratory at a much faster rate by providing a harsher, but nonetheless representative, environment. In such a test, the product is expected to fail in the lab just as it would have failed in the field—but in much less time. The main objective of an accelerated test is either of the following:

To discover failure modes
To predict the normal field life from the high-stress lab life

Accelerated testing needs planning as follows:

Define the objective and scope of the test.
Collect required information about the product.
Identify the stress(es).
Determine the level of stress(es).
Conduct the accelerated test, and analyze the accelerated data.

Common ways to determine a life stress relationship are the following:

Arrhenius model
Eyring model
Inverse power law model
Temperature-humidity model
Temperature nonthermal model

Software Reliability

Software reliability is a special aspect of reliability engineering. System reliability, by definition, includes all parts of the system, including hardware, software, operators, and procedures. Traditionally, reliability engineering focuses on critical hardware parts of the system. Since the widespread use of digital integrated circuit technology, software has become an increasingly critical part of most electronics and, hence, nearly all present-day systems. There are significant differences, however, in how software and hardware behave. Most hardware unreliability is the result of a component or material failure that results in the system not performing its intended function. Repairing or replacing the hardware component restores the system to its original unfailed state. However, software does not fail in the same sense that hardware fails. Instead, software unreliability is the result of unanticipated results of software operations. Even relatively small software programs can have astronomically large combinations of inputs and states that are infeasible to exhaustively test. Restoring software to its original state only works until the same combination of inputs and states results in the same unintended result. Software reliability engineering must take this into account.

Despite this difference in the source of failure between software and hardware—software doesn't wear out—some in the software reliability–engineering community believe statistical models used in hardware reliability are nevertheless useful as a measure of software reliability, describing what we experience with software: the longer you run software, the higher the probability you'll eventually use it in an untested manner and find a latent defect that results in a failure (Shooman 1987; Musa 2005; Denney 2005).

As with hardware, software reliability depends on good requirements, design, and implementation. Software reliability engineering relies heavily on a disciplined software-engineering process to anticipate and design against unintended consequences. There is more overlap between software quality engineering and software reliability engineering than between hardware quality and reliability. A good software development plan is a key aspect of the software reliability program. The software development plan describes the design and coding standards, peer reviews, unit tests, configuration management, software metrics, and software models to be used during software development.

A common reliability metric is the number of software faults, usually expressed as faults per thousand lines of code. This metric, along with software execution time, is key to most software reliability models and estimates. The theory is that the software reliability increases as the number of faults (or fault density) goes down. Establishing a direct connection between fault density and MTBF is difficult, however, because of the way software faults are distributed in the code, their severity, and the probability of the combination of inputs necessary to encounter the fault. Nevertheless, fault density serves as a useful indicator for the reliability engineer. Other software metrics, such as complexity, are also used.

Testing is even more important for software than hardware. Even the best software development process results in some software faults that are nearly undetectable until tested. As with hardware, software is tested at several levels, starting with individual units, through integration and full-up system testing. Unlike with hardware, it is inadvisable to skip levels of software testing. During all phases of testing, software faults are discovered, corrected, and retested. Reliability estimates are updated based on the fault density and other metrics. At the system level, MTBF data are collected and used to estimate reliability. Unlike with hardware, performing the exact same test on the exact same software configuration does not provide increased statistical confidence. Instead, software reliability uses different metrics such as test coverage.

Eventually, the software is integrated with the hardware in the top-level system, and software reliability is subsumed by system reliability. The Software Engineering Institute's *capability maturity model* is a common means of assessing the overall software development process for reliability and quality purposes.

Reliability Operational Assessment

After a system is produced, reliability engineering during the system operation phase monitors, assesses, and corrects deficiencies. Data collection and analysis are the primary tools used. When possible, system failures and corrective actions are reported to the reliability engineering organization. The data are constantly analyzed using statistical techniques, such as Weibull analysis and linear regression, to ensure the system reliability meets the specification. Reliability data and estimates are also key inputs for system logistics. Data collection is highly dependent on the nature of the system. Most large organizations have quality control groups that collect failure data on vehicles, equipment, and machinery. Consumer product failures are often tracked by the number of returns. For systems in dormant storage or on standby, it is necessary to establish a formal surveillance program to inspect and test random samples. Any changes to the system, such as field upgrades or recall repairs, require additional reliability testing to ensure the reliability of the modification.

Reliability Organizations

Systems of any significant complexity are developed by organizations of people, such as a commercial company or a government agency. The reliability-engineering organization must be consistent with the company's organizational structure. For small, noncritical systems, reliability engineering may be informal. As complexity grows, the need arises for a formal reliability function. Because reliability is important to the customer, the customer may even specify certain aspects of the reliability organization.

There are several common types of reliability organizations. The project manager or chief engineer may employ one or more reliability engineers directly. In larger organizations, there is usually a product assurance or specialty-engineering organization, which may include reliability, maintainability, quality, safety, human factors, logistics, and so on. In such case, the reliability engineer reports to the product assurance manager or specialty-engineering manager.

In some cases, a company may wish to establish an independent reliability organization. This is desirable to ensure that the system reliability, testing of which is often expensive and time-consuming, is not unduly slighted due to budget and schedule pressures. In such cases, the reliability engineer works on the project on a day-to-day basis, but is actually employed and paid by a separate organization within the company.

Because reliability engineering is critical to early system design, it has become common for reliability engineers; however, the organization is structured to work as part of an integrated product team.

Certification

The American Society for Quality (ASQ) has a program to become a certified reliability engineer, or CRE. Certification is based on education, experience, and a certification test; periodic recertification is required. The body of knowledge for the test includes reliability management, design evaluation, product safety, statistical tools, design and development, modeling, reliability testing, collecting and using data, and so on.

Reliability Engineering Education

Some universities offer graduate degrees in reliability engineering (e.g., the University of Maryland). Reliability engineers typically have an engineering degree, which can be in any field of engineering, from an accredited university or college program. Many engineering programs offer reliability courses, and some universities have entire reliability-engineering programs. A reliability engineer may be registered as a professional engineer by the state, but this is not required by most employers. There are many professional conferences and industry training programs available for reliability engineers. Several professional organizations exist for reliability engineers, including the IEEE Reliability Society, the ASQ, and the Society of Reliability Engineers (SRE).

SYSTEMS ANALYSIS

System analysis is the branch of electrical engineering that characterizes electrical systems and their properties. Although many of the methods of system analysis can be applied to nonelectrical systems, it is a subject often studied by electrical engineers because it has direct relevance to many other areas of their discipline, most notably signal processing and communication systems.

Characterization of Systems

A system is characterized by how it responds to input signals. In general, a system has one or more input signals and one or more output signals. Therefore, one natural characterization of systems is by how many inputs and outputs they have:

Single input, single output (SISO)
Single input, multiple outputs (SIMO)
Multiple inputs, single output (MISO)
Multiple inputs, multiple outputs (MIMO)

It is often useful (or necessary) to break up a system into smaller pieces for analysis. Therefore, we can regard a SIMO system as multiple SISO systems (one for each output), and the same applies for a MIMO system. By far, the greatest amount of work in system analysis has been with SISO systems, although many parts inside SISO systems have multiple inputs (such as adders).

Signals can be continuous or discrete in time, as well as continuous or discrete in the values they take at any given time:

- Signals that are continuous in time and continuous in value are known as *analog signals.*
- Signals that are discrete in time and discrete in value are known as *digital signals.*
- Signals that are discrete in time and continuous in value are called *discrete time signals.* While important mathematically, systems that process discrete time signals are difficult to physically realize. The methods developed for analyzing discrete time signals and systems are usually applied to digital and analog signals and systems.
- Signals that are continuous in time and discrete in value are sometimes seen in the timing analysis of logic circuits, but have little to no use in system analysis.

With this categorization of signals, a system can then be characterized as to which type of signals it deals with:

- A system that has analog input and analog output is known as an *analog system.*
- A system that has digital input and digital output is known as a *digital system.*

Systems with analog input and digital output or digital input and analog output are possible. However, it is usually easiest to break up these systems into their analog and digital parts for analysis, as well as the necessary analog-to-digital or digital-to-analog converter.

Another way to characterize systems is by whether their output at any given time depends only on the input at that time, or perhaps on the input at some time in the past (or in the future!).

Memoryless systems do not depend on any past input.
Systems with memory do depend on past input.
Causal systems do not depend on any future input.
Noncausal or anticipatory systems do depend on future input.

Note: It is not possible to physically realize a noncausal system operating in "real time." However, from the standpoint of analysis, these systems are important for two reasons. First, the ideal system for a given application is often a noncausal system, which although not physically possible, can give insight into the design of a derivated causal system to accomplish a similar purpose. Second, there are instances when a system does not operate in "real time" but rather is simulated "offline" by a computer.

Analog systems with memory may be further classified as lumped or distributed. The difference can be explained by considering the meaning of memory in a system. Future output of a system with memory depends on future input and a number of state variables, such as values of the input or output at various times in the past. If the number of state variables necessary to describe future output is finite, the system is lumped; if it is infinite, the system is distributed.

Finally, systems may be characterized by certain properties which facilitate their analysis:

A system is linear if it has superposition and scaling properties.

A system that is not linear is nonlinear.

If the output of a system does not depend explicitly on time, the system is said to be time-invariant; otherwise, it is time-variant,

A system that will always produce the same output for a given input is said to be deterministic.

A system that will produce different outputs for a given input is said to be stochastic.

There are many methods of analysis developed specifically for linear time-invariant (LTI) deterministic systems. Unfortunately, in the case of analog systems, none of these properties are ever perfectly achieved. Linearity implies that operation of a system can be scaled to arbitrarily large magnitudes, which is not possible. Time-invariance is violated by aging effects that can change the outputs of analog systems over time (usually years or even decades). Thermal noise and other random phenomena ensure that the operation of any analog system will have some degree of stochastic behavior. Despite these limitations, however, it is usually reasonable to assume that deviations from these ideals will be small.

LTI Systems

As mentioned above, there are many methods of analysis developed specifically for LTI systems. This is due to their simplicity of specification. An LTI system is completely specified by its transfer function (which is a rational function for digital and lumped analog LTI systems). Alternatively, we can think of an LTI system as being completely specified by its frequency response. A third way to specify an LTI system is by its characteristic linear differential equation (for analog systems) or linear difference equation (for digital systems). Which description is most useful depends on the application.

The distinction between lumped and distributed LTI systems is important. A lumped LTI system is specified by a finite number of parameters, be it the zeros and poles of its transfer function, or the coefficients of its differential equation, whereas specification of a distributed LTI system requires a complete function.

AUDITING

Quality audit is the process of systematic examination of a quality system carried out by an internal or external quality auditor or an audit team. It is an important part of an organization's quality management system and is a key element in the ISO quality system standard, ISO 9001.

Quality audits are typically performed at predefined time intervals and ensure that the institution has clearly defined internal quality-monitoring procedures linked to effective action. This can help determine if the organization complies with the defined quality system processes and can involve procedural or results-based assessment criteria.

With the upgrade of the ISO 9000 series of standards from the 1994 to 2000 series, the focus of the audits has shifted from purely procedural adherence toward measurement of the actual effectiveness of the quality management system (QMS) and the results that have been achieved through the implementation of a QMS.

Quality audits can be an integral part of compliance for regulatory requirements. One example is the U.S. Food and Drug Administration, which requires quality auditing to be performed as part of its Quality System Regulation (QSR) for medical devices (Title 21 of the U.S. Code of Federal Regulations, part 820).

Several countries have adopted quality audits in their higher education system (New Zealand, Australia, Sweden, Finland, Norway, and the United States). Initiated in the UK, the process of quality audit in the education system focused primarily on procedural issues rather than on the results or the efficiency of a quality system implementation.

Audits can also be used for safety purposes. Evans and Parker (2008) describe auditing as one of the most powerful safety-monitoring techniques and "an effective way to avoid complacency and highlight slowly deteriorating conditions," especially when the auditing focuses not just on compliance but also on effectiveness.

Audit Planning and Scheduling

Auditor Education and Training

An auditor must possess and maintain sufficient basic education and training in order to perform audits in a professional manner.

Audit Initiation

Audits are initiated by the client either by request or through approval of a program of audits submitted by the auditing department or group. The audit must be assigned to and be accepted by a qualified auditor.

Audit Scope

The scope of audits depends on the need as determined by the client and/or auditing organization. In most cases, the scope of the quality system is defined in the top-level quality manual.

Audit Objective

Audits determine compliance or noncompliance with established standards and assess the effectiveness of such standards. The secondary objective of a quality audit can be to determine opportunities and needs for improvements in the operation and control systems, review performances and results, and facilitate communication. The intent of an audit is that the auditor obtains sufficient evidence to draw conclusions relative to the stated audit objective.

PLAN, SCHEDULE, AND RESULTS

	1. Audit plan	2. Schedule		3. Report (Result)			
				Conformed?			
Element	Auditor(s) Assigned	Date(s) Audited	Next Audit	N/A	Yes	No	AR
4.22 Quality manual				☐	☐	☐	
4.2.3 Control of documents				☐	☐	☐	
4.2.4 Control of records				☐	☐	☐	
5.1 Management commitment				☐	☐	☐	
5.2 Customer focus				☐	☐	☐	
5.3 Quality focus				☐	☐	☐	
5.4.1 Quality objectives				☐	☐	☐	
5.4.2 Quality planning				☐	☐	☐	
5.5.1 Responsibility and authority				☐	☐	☐	
5.5.3 Internal communications				☐	☐	☐	
5.6 Management review				☐	☐	☐	
6.2.2 Competence awareness and training				☐	☐	☐	
6.3 Infrastructure				☐	☐	☐	
7.2 Customer related process				☐	☐	☐	
7.3 Design and development				☐	☐	☐	
7.4 Purchasing				☐	☐	☐	
7.5.1 Production provision				☐	☐	☐	
7.5.2 Validation of process				☐	☐	☐	
7.5.3 Indentification and traceablity				☐	☐	☐	
7.5.4.Customer property				☐	☐	☐	
7.5.5 Preservation of product				☐	☐	☐	
7.6 Calibration				☐	☐	☐	
8.2.1 Customer satisfaction				☐	☐	☐	
8.2.2 Internal audit				☐	☐	☐	
8.2.3 Monitoring and measurement of processes				☐	☐	☐	
8.2.4 Monitoring and measurement of product				☐	☐	☐	
8.4 Analysis of data				☐	☐	☐	
8.5.3. Preventative action				☐	☐	☐	
				☐	☐	☐	

COMMENTS & OBSERVATIONS

FIGURE 6.2
Audit plan and report.

Frequency and Timing

Audit frequency may be determined by law or regulation, by the audit program, by standards, or by the need of the client.

The timing should be chosen with due regard to availability of evidential material, unbiased observations, adequate cooperation and support from the auditee, sufficiency of the audit resources, and least cost.

Long-term planning provides a framework for an annual audit program. The individual audit assignments in the program must be further planned in detail.

Long-Term Planning

This is usually carried out by the audit department or group. The resulting plan or program (see Figure 6.2) should be approved by the client. The plan, or program, should include the name of the organizational unit, the object of the audit, and the expected duration and timing of each audit element.

Pre-Audit Review of System

Audits should be planned and carried out only where a quality system is established. Pre-audit reviews are to verify the existence of a system or individual documented procedure that can be audited.

The planning should be conducted by the auditor, or lead auditor with the assistance of the auditors assigned to the team. Audit elements assigned to the individual auditors should be coordinated and integrated in the audit plan.

Working Papers

These are all of the documents required for an effective and orderly execution of the audit plan (see Figure 6.3).

Result

Sampling Plans

Sampling plans are used in the audit to ensure applicability, validity, and reliability of the observation being made (see Figure 6.4).

AUDITOR	DATE:	
Activity	Comments	Rating
Quality management system 4.1. General requirements The organization shall establish, document, implement, and maintain a quality management system and continually improve its effectiveness in accordance with the requirements of this International Standard. The organization shall a) Identify the processes needed for the quality management system and their application throughout organization (see 12). b) Determine the sequence and interaction of these processes. c) Determine criteria and methods needed to ensure that both the operation and control of these processes are effective. d) Ensure the availibility of resources and information necessary to support the operation and monitoring of these processes. e) Monitor, measure and analyze these processes. f) Implement actions necessary to achieve planned results and continual improvement of these processes. These processes shall be managed by the organization in accordance with the requirements of this International Standard. When an organization chooses to outsource any process that affects product conformity with requirements, the organization shall ensure control over such processes. Control of such outsource processes shall be identified within the quality management system. Note: Process needed for the quality mangement system referred to above should include processes for management activities.		

FIGURE 6.3

Audit working paper.

C=0 SAMPLING PLAN

This table is read starting at the left-hand column, reading down and to the right, and finding the correct sample size under the appropriate AQL. The lot is rejected if one non-conformance is found.

LOT SIZE		AQLs/SAMPLE SIZES			
FROM	TO	2.2	4	6.5	10
2	8	5	3	2	2
9	15	5	3	2	2
16	25	5	3	3	2
26	50	5	5	5	3
51	90	7	6	5	4
91	150	11	7	6	5
151	280	13	10	7	6
281	500	16	11	9	7
501	1,200	19	15	11	8
1,201	3,200	23	18	13	9
3,201	10,000	29	22	15	9
10,001	35,000	35	29	15	9
35,001	150,000	40	29	15	9
150,001	500,000	40	29	15	9
500,001	>500,001	40	29	15	9

FIGURE 6.4
C = 0 sampling plan.

Audit Implementation Steps

The audit plan should be implemented through the following steps:

Notification to the auditee
Orientation of auditors and auditee
Examination
Follow-up and close-out
Reporting of results (management review)

Notification to Auditee

Advance notification allows the auditee to make final preparations for the audit. The audit plan should be forwarded with the notification.

Opening Meeting

The audit team should meet when final preparation and decisions need to be made. A brief meeting with the management of the organization to be audited serves for clarification of the audit plan, introduction of the auditors, and finalization of procedures and meetings.

Information, Verification, and Evaluation

The auditor must obtain sufficient, relevant information and evidence that permit a valid and reliable verification and evaluation (see Figure 6.5).

Audit Observations

Audit observations are significant conclusions and results of the examination.

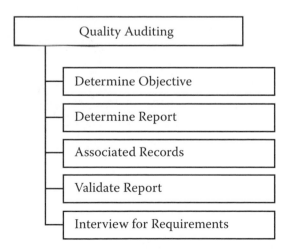

FIGURE 6.5
Audit steps: Any process.

Audit Supervision

At the conclusion of the audit and prior to preparing the audit report, a meeting should be held by the auditor or lead auditor with the auditee's senior management. The main purpose of the meeting is to present and clarify all audit observations to be reported, along with supporting evidence, so that the auditee can initiate necessary corrective action effectively without delay.

Audit Follow-Up

Follow-up consists of verification of corrective action resulting from observations.

Preparation of the Report

Standards for the form and content of the report should be established and followed.

Content of the Report

The audit report should include the following:

> Purpose, objective, and scope of the audit
> Details of the audit plan, auditors, dates, and organization audited
> Standards used
> Observations and evidence
> Noteworthy comments and recommendations
> Follow-up corrective actions

Reporting the Audit

Review and Distribution

Management of the auditing organization should review and approve the report prior to submitting it to the client. The client decides on the distribution of copies of the report.

Audit Completion

An audit assignment is completed upon submission of the audit report to the client, except in special circumstances when verification of corrective action is explicitly included in the audit assignment and plan.

Record Retention

The auditor, lead auditor, or audit organization is responsible for custody and retention of audit documents and records.

COST OF QUALITY

In management accounting, cost accounting is the process of tracking, recording, and analyzing costs associated with the products or activities of an organization. Managers use cost accounting to support decision making to reduce a company's costs and improve its profitability. As a form of management accounting, cost accounting need not follow standards such as generally accepted accounting principles (GAAP), because its primary use is for internal managers, rather than external users, and what to compute is instead decided pragmatically.

Costs are measured in units of nominal currency by convention. Cost accounting can be viewed as translating the supply chain (the series of events in the production process that, in concert, result in a product) into financial values.

There are at least four approaches:

Standardized cost accounting
Activity-based costing
Throughput accounting
Marginal costing, or cost-volume-profit analysis

Classical cost elements are as follows:

Raw materials
Labor
Allocated overhead

TABLE 6.6

Cost-of-Quality Statement

		Cost	Subtotal
Failure Costs			
	Raw material nonconformance	$3,276	
	Repairs	$70,299	
	Scrap	$2,000	
	Returns	$300,000	
	Rework	$20,000	
			$395,575
Appraisal Costs			
	Product audits	$32,000	
	Receiving inspection	$25,000	
	In-process inspection	$25,000	
	Final inspection	$50,000	
			$132,000
Prevention Costs			
	Design reviews	$9,000	
	Quality assurance	$25,000	
			$34,000
Total			$561,575

Origins

Cost accounting has long been used to help managers understand the costs of running a business. Modern cost accounting originated during the Industrial Revolution, when the complexities of running a large-scale business led to the development of systems for recording and tracking costs to help business owners and managers make decisions.

In the early industrial age, most of the costs incurred by a business were what modern accountants call *variable costs* because they varied directly with the amount of production. Money was spent on labor, raw materials, power to run a factory, and so on, in direct proportion to production. Managers could simply total the variable costs for a product and use this as a rough guide for decision-making processes.

Some costs tend to remain the same even during busy periods, unlike variable costs, which rise and fall with volume of work. Over time, the importance of these "fixed costs" has become more important to managers.

Examples of fixed costs include the depreciation of plant and equipment, and the cost of departments such as maintenance, tooling, production control, purchasing, quality control, storage and handling, plant supervision, and engineering. In the early twentieth century, these costs were of little importance to most businesses. However, in the twenty-first century, these costs are often more important than the variable cost of a product, and allocating them to a broad range of products can lead to bad decision making. Managers must understand fixed costs in order to make decisions about products and pricing.

The concept of quality costs (see Table 6.6) is a means to quantify the total cost of quality-related efforts and deficiencies. It was first described by Armand V. Feigenbaum in a 1956 *Harvard Business Review* article.

Prior to its introduction, the general perception was that higher quality requires higher costs, either by buying better materials or machines, or by hiring more labor. Furthermore, while cost accounting had evolved to categorize financial transactions into revenues, expenses, and changes in shareholder equity, it had not attempted to categorize costs relevant to quality. By classifying quality-related entries from a company's general ledger, management and quality managers can evaluate investments in quality based on cost improvement and profit enhancement.

Internal failure costs: These are costs associated with nonconformities that are found during receiving, in-process inspection, and finished-goods inventory prior to shipping to the customer. Examples would be the following:

Scrap
Rework
Supplier scrap or rework
Sorting
Retest and reinspection
Regrading

External failure costs: These are costs associated with nonconformities that are found by the customer. Examples would be the following:

Warranty charges
Service time and material allowances
Returned material costs

Appraisal costs: These are the costs associated with product verification and validation. Examples would be as follows:

Product audits
Receiving inspection
In-process inspection
Final inspection
Calibration

Prevention costs: These are the costs associated with activities associated with preventing nonconformities from occurring. Examples would be as follows:

Planning
Reviews
Management controls
Organizing
Internal audits
Supplier audits
Training

Appendix A: Business Quality Plan

Organization				Objective and Goals	
1. Categories and Classifications	2. Department	3. Responsibility		4. Tracking	5. Goal or Objective
		Primary	Alternate		

Appendix B: Process Quality Plan

General Information			
No.	Classification (Process):		Date:
Phase: ☐Design ☐Review ☐Production		Contact Name:	Phone:
Department:	Primary:		Alternate:
Tracking:		Goal:	

1. Flowchart					2. Process Step Description	3. Requirement (Product or Process)	4. Possible Problems	5. Possible Causes	6. Sensor	7. Methods		8. Document	9. Reaction Plan
Operation	Transportation	Inspection	Delay	Storage						Sample	Frequency		
Receiving (Input)													
O	O	O	O	O									
O	O	O	O	O									
O	O	O	O	O									
Process (WIP)													
O	O	O	O	O									
O	O	O	O	O									
O	O	O	O	O									
O	O	O	O	O									
O	O	O	O	O									
O	O	O	O	O									
O	O	O	O	O									
O	O	O	O	O									
Final (Output)													
O	O	O	O	O									
O	O	O	O	O									
O	O	O	O	O									

Appendix C: Product Quality Plan

Inspection Plan For:		Effective Date:	Number:
		Supersedes:	Page: of
Approved By:			

Instructions:
Inspect the product with regard to the characteristics listed below (also see inspection and test work instructions). Use a C = 0 sampling plan with an AQL of 10 unless otherwise specified below. Record the results of the inspection on the appropriate inspection report or log. In the event of a nonconformity, follow work instruction.

No.	Characteristics to Be Measured or Inspected	Specification and Tolerance (±)	AQL	Inspection or Measuring Equipment or Method	Comments
1					
2					
3					
4					
5					
6					
7					
8					
9					
10					
11					
12					
13					
14					
15					
16					
17					
18					
19					
20					

21					
22					
23					
24					
25					
26					
28					
29					

Bibliography

Byrns, R. T., and G. W. Stone. *Macro Economics*, 2nd ed. Glenview, IL: Scott, Foresman, 1995.

Corley, R. N., E. M. Holmes, and W. J. Robert. *Fundamentals of Business Law*, 3rd ed. Englewood Cliffs, NJ: Prentice Hall, 1982.

Dessler, G. *Personnel Management*, 4th ed. Englewood Cliffs, NJ: Prentice Hall, 1988.

Gore, M., and J. Stubbe. *Elements of System Analysis*, 3rd ed. Dubuque, IA: Brown, 1983.

Grant, E. L., and R. S. Leavenworth. *Statistical Process Control*, 5th ed. New York: McGraw-Hill, 1980.

Hayslett, II. T., Jr. *Statistics Made Simple*. New York: Doubleday, 1967.

Hersey, P., and K. Blanchard. *Management of Organizational Behavior*, 4th ed. Englewood Cliffs, NJ: Prentice Hall, 1982.

Institute of Electrical and Electronics Engineers. *IEEE 1332: Standard Reliability Program for the Development and Production of Electronic Systems and Equipment*. June 30. Piscataway, NJ: IEEE, 1998.

Juran, J. M. *Quality Control Handbook*. New York: McGraw-Hill, 1951.

Knowles, M. S. *The Adult Learner: A Neglected Species*. Houston, TX: Gulf, 1990.

Kolin, P. C. *Successful Writing at Work*. Lexington, MA: D. C. Heath, 2003.

Laird, D., and P. R. Schleger. *Approaches to Training and Development*. New York: Perseus, 1985.

Levin, R. I., C. A. Kirkpatrick, and D. S. Rubin. *Quantitative Approaches to Management*. New York: McGraw-Hill, 1982.

Mandell, R. L., S. L. Cowen, and S. S. Miller. *Introduction to Business: Concepts and Applications*. St. Paul, MN: West, 1981.

Messerschmitt, D. G., and C. Szyperski. Marketplace Issues in Software Planning and Design. *IEEE Software* 21, no. 3 (May–June 2004): 62–70.

Mills, C. A. *The Quality Audit*. New York: McGraw-Hill, 1989.

Moriarity, S., and C. P. Allen. *Cost Accounting*. New York: Harper & Row, 1987.

Neeley, L. P., and F. J. Imke. *Accounting Principles and Practices*, 2nd ed. Cincinnati, OH: South-Western, 1987.

O'Connor, P. D. T. *Practical Reliability Engineering*. New York: Wiley, 1981.

Rue, L. W., and L. L. Byars. *Management Theory and Application*, 3rd ed. Homewood, IL: Irwin, 1983.

Runyon, R., P., and A. Haber. *Business Statistics*. Homewood, IL: Irwin, 1982.

Stanton, W. J., and C. M. Futrell. *Fundamentals of Marketing*, 8th ed. New York: McGraw-Hill, 1986.

U.S. Air Force. *MIL-STD-785B (Notice 1), Military Standard, Reliability Program for Systems and Equipment Development and Production*. July 3, 1986. http://www.everyspec.com/MIL-STD/MIL-STD+(0700+-+0799)/download.php?spec=MIL_STD_785B_NOT_1.1010.pdf (last accessed August 3, 2009).

Wikipedia. *Wikipedia: The Free Encyclopedia*. http://www.wikipedia.org (last accessed July 30, 2009).

Index

process control, 97, 102–104,
106
sample, 94
theory, 95
time series, 94
variation types, 103
Support center
classifications of, 6
responsibilities of, 44
role of, 4
Systems analysis
characterization of systems,
122–124
linear time-invariant (LTI),
124, 125
signals, 123
Systems approach to management, 40

T

Taylor, Frederick, 40
Theory, definition of, 64

Training
overview, 73–74
records, maintaining, 74, 75*f*

U

Unity of command, 9

V

Validation planning
facilities for validation, 29
overview, 28
personnel, 30
procedures, 30
revisions to, 30
scheduling, 30
Value
importance of, 42–43
quality reporting, as part of, 47
Variable, definition of, 1
Vocational rehabilitation, 67